The Reflector

The Reflector

Jennifer Battaglia, MS
Margie Hodges Shaw, JD, PhD
Susan Daiss, MA, MDiv
Martha J. Gdowski, PhD

MELIORA PRESS

An imprint of the University of Rochester Press

First published 2021

Meliora Press is an imprint of the
University of Rochester Press
668 Mt. Hope Avenue, Rochester, NY 14620, USA
www.urpress.com
and Boydell & Brewer Limited
PO Box 9, Woodbridge, Suffolk IP12 3DF, UK
www.boydellandbrewer.com

ISBN-13: 978-1-64825-013-2

Library of Congress Cataloging-in-Publication Data

Names: Battaglia, Jennifer, author.
Title: The reflector / Jennifer Battaglia, Margie Hodges Shaw, Susan Daiss, Martha J.
    Gdowski.
Description: Rochester : Meliora Press, 2021. | Includes bibliographical references.
Identifiers: LCCN 2020056214 | ISBN 9781648250132 (paperback)
Subjects: LCSH: Human dissection.
Classification: LCC QM33.5 .B38 2021 | DDC 611—dc23
LC record available at https://lccn.loc.gov/2020056214

This publication is printed on acid-free paper.
Printed in the United States of America.

Title page image by Jennifer Hu, MD, University of Rochester School of Medicine and
Dentistry, Class of 2018.

CONTENTS

Preface   vii

A Daughter's Letter to Anatomy Students   ix

Introduction   1

The Humanity of Anatomy   3

Block I

Chapter 1: Back and Spinal Cord   7
  I. Back   7
  II. Deeper Back Muscles and Spinal Cord   9

Chapter 2: Upper Limb   13
  I. Shoulder   13
  II. Axilla and Arm   17
  III. Flexor Forearm   21
  IV. Palm   23
  V. Extensor Forearm and Dorsal Hand   26
  VI. Joints of Upper Limb   28

Block I Bold Terms   29
Block I Reflections   37

Block II

Chapter 3: Thorax and Root of the Neck   41
  I. Thoracic Wall   41
  II. Pleural Cavities and the Lungs   48
  III. Mediastinum and Heart   49
  IV. Posterior and Superior Mediastinum   55
  V. Anterior Triangle and Root of the Neck   57
  VI. Posterior Triangle of the Neck   62

Block II Bold Terms   63
Block II Reflections   69

Block III

Chapter 4: Head and Neck   73
  I. Face   73
  II. Interior of Skull and Brain Removal   76

III. Orbit and Eye   85
IV. Temporal Region   88
V. Retropharyngeal Region   90
VI. Pharynx   92
VII. Tongue and Nasal Cavity   93
VIII. Palate, Mouth, and Nasopharyngeal Wall   96
IX. Larynx   98
X. Ear   100

Block III Bold Terms   105
Block III Reflections   115

Block IV

Chapter 5: Abdomen   119
I. Anterior Abdominal Wall   119
II. Scrotum, Spermatic Cord, and Testis   120
III. Peritoneum and Peritoneal Cavity   124
IV. Bile Passages, Celiac Trunk, and Portal Vein   128
V. Superior and Inferior Mesenteric Vessels   131
VI. Removal of GI Tract   132

Block IV Bold Terms   135
Block IV Reflections   141

Block V

Chapter 6: Posterior Abdomen   145
I. Posterior Abdominal Structures   145
II. Posterior Abdominal Wall   150

Chapter 7: Pelvis and Perineum   153
I. Pelvis   153
II. Transection of Abdomen and Splitting of Pelvis   157
III. Urogenital Triangle   157

Chapter 8: Lower Limb   161
I. Anterior and Medial Thigh   161
II. Gluteal Region and Posterior Thigh   166
III. Popliteal Fossa and Leg   167
IV. Sole of the Foot   174
V. Joints of Lower Limb   175

Block V Bold Terms   177
Block V Reflections   185

Post-Lab Pause   187

Sources   189

PREFACE

The donors in an anatomy lab have gifted to you the most valuable thing that they had at the conclusion of their lives to allow you to learn from them. Acknowledgment of that gift can feel like a heavy burden. Yet, the permission to learn as much as you can through their dissection is an implicit message to you from this donor that reinforces the value of your education and the difference that you will make in the medical profession as a consequence of your training.

You will never know your donor in the way you may know the lab partners with whom you share this dissection experience. You will never be able to ask the questions of who and what they loved in life, what they regretted, whether they had a good life and a good death, or why they chose to gift their body. Yet, you will come to know them in a memorable way. When you auscultate your patients' aortic and pulmonic valves and superior, middle, or inferior lobes of the lungs in the clinic, it will be your donor's heart and lungs that you visualize. Without ever knowing you, your donor had insight to entrust you to learn from them through this unique experience.

Virtual anatomy dissection apps simulate the systematic removal of tissue layers to visualize deeper structures. Photographic videos reveal anatomical structures evident at various stages of dissection. These tools facilitate recall of useful landmarks for identifying anatomical structures but won't cultivate the motor memory achieved by separation of fascial planes and reflection of structures to reveal what lies beneath. They won't force you to find ways to work as a team through a dissection that is emotionally or physically challenging. Seemingly superfluous to the dissection itself, that process of learning to navigate difficulty with your colleagues lays a foundation for challenges encountered in future care of patients.

Through the years, senior students have shared with me that anatomy lab was among the most stressful parts of their medical education. Additional queries unveiled two main stressors. The first was that team dysfunction hindered learning. I encourage you to get to know each other outside of the anatomy laboratory. Share your passions and dislikes, strengths and weaknesses in learning, and individual goals for the anatomy lab. Then, work as a team to formulate a plan that allows each of you to be successful in that space in your own ways, all while functioning as a team and supporting each other in your collective learning.

The second stressor was related to human dissection itself. As an instructor, I know that there are psychologically challenging dissections. The emotional valence to the hand, for example, makes the dissection of the palm more challenging than other regions. Losses of friends or family related to diseases in specific body regions may trigger unexpected reactions to specific dissections. While the dissections support the acquisition of anatomy knowledge expected of clinicians, your reactions to specific challenges in the anatomy lab establish patterns of response to challenges encountered in clinical practice. This *Reflector* was created by a medical student with the intention of supporting your successful navigation of challenges and formulation of effective tools for evaluating those experiences during dissection and afterwards.

You will be changed by your experience in the anatomy lab. For some, that change will be palpable and welcomed, for others, it will be accepted with reluctance and angst. Embrace that everyone will process this experience differently. Periodically contemplate events in the anatomy lab that were transformational. I see those as opportunities for great personal and professional growth and I challenge you to use them as such.

—Martha J. Gdowski, PhD in Anatomy from The Pennsylvania State University College of Medicine, Anatomical Sciences Strand Director for Human Structure and Function at the University of Rochester School of Medicine, Course Director and Sole Instructor for Human Anatomy and Applied Human Anatomy.

A DAUGHTER'S LETTER TO ANATOMY STUDENTS

July 1997

Dear University of Rochester Students,

Have you ever wandered into an uninhabited old house and wondered what life was like when its first residents and those who followed them lived there? Have you ever tried to picture it brimming over with children playing, filled with smells from the kitchen and the sound of music from a piano in the front parlor? Or, have you ever sat in someone else's living room waiting for them to return from the kitchen with a cup of tea and tried to figure them out based on the pictures on the wall, the knick-knacks on a bookshelf, or the crocheting sitting next to a favorite chair? If you have had these or similar experiences, then perhaps you can appreciate a little more what lies before you in anatomy.

As you enter the anatomy lab and work on a donor's body, imagine that you are wandering into a house that could easily have belonged to my Dad, who has died already and donated his body, or could one day belong to my Mom, who has plans to donate hers.

People who know my Mom and Dad's house enter through the back door. You, too, will enter through the rear part of the person you are dissecting . . .

- A back not unlike that which held me as a child when I wrapped my arms around my Dad's neck as he waded out into the water at our favorite vacation spot in the Adirondacks.
- A back bent with age and arthritis that must have caused pain and suffering to someone like my Mom, who would like to be as active as she once was, bringing communion to those in nursing homes or bending down to pick up a crying child from the pavement.

As you eventually hold this person's brain in your hands it is as though you have been invited to see an upstairs room . . .

A brain as good as yours, perhaps, which could have brought this person to the University of Rochester to study chemistry if it were not for the Depression in the 1930s and the need to get a job in the Civilian Conservation Corps to support the family.

The brain of one who loves Lawrence Welk's music, hates Frank Sinatra's, enjoys singing in the church choir, and whose greatest desire was to go to Toronto to see the stage musical *Ragtime.*

As you carefully enter into nooks and crannies of the hands, you will spend more time exploring these than it would take to explore my parents' whole house . . .

- Hands that may have held the back of a bike until a little girl could ride on her own or that carried suitcases when that little girl grew up and went off to study, to work, or to see the world and experience new places.
- Hands that may have baked the best strawberry pie going, whipped up a roast beef and Yorkshire pudding for Sunday dinner, and then flown over the piano keys as they played "Edelweiss Glide" or "Nola."

Each room is before you to explore with care the texture, the content, the color, the defects, and the beauty of the human form. Enter each room with awe because it holds a cherished memory in the hearts of those who knew the first inhabitant.

Peace on your journey of discovery,

—The Daughter of Anatomical Gift Donors

INTRODUCTION

Purpose

Cadaver: Latin = one who has fallen.

The purpose of this book is to serve as a reminder that medicine is not only a science, but also an art of understanding how our own individuality interacts with the individuality of those around us. I see *The Reflector* as a humanities-based tool to help students learn anatomy, and to begin to establish the balance between emotional responses and objective responsibilities. In order to grow personally and professionally we, as students, need time and space to think about our values, beliefs, and experiences and recognize that these elements will mold us into the physicians that we will become.

About *The Reflector*: A Guide for Students

The Reflector parallels the dissection manual used at the University of Rochester School of Medicine and Dentistry, titled *The Dissector*, but the format can be adapted to any anatomy curriculum. Ultimately this book is for you—you are free to use it however you'd like. *The Reflector* is meant to support a wide range of studies, from anatomical to professional to personal. It is my hope that students with different academic backgrounds and with different learning styles can use this manual as an aid throughout the duration of anatomy lab. You also may want to revisit this manual if you return to anatomy lab for additional coursework in the future.

This book is only a foundation for the many thoughts, questions, and emotions that you will encounter in the anatomy lab and throughout medical school. Never hesitate to talk with one another; I challenge you to broaden your perspectives and to learn from each other's experiences, strategies, and beliefs.

Many of the images in *The Reflector* are by Emily Evans and are reproduced from *The Secret Language of Anatomy*, by C. Brassett, E. Evans, and I. Fay (2017). We encourage students to purchase their own copy of the book. The images from *The Secret Language of Anatomy* are reproduced with the generous permission of Lotus Publishing. You can find out more about *The*

Secret Language of Anatomy at the publisher's website and it is available for purchase in the United States from a variety of sources, including Amazon. Finally, the names of the donors in student reflections were changed to protect the privacy of those who donated their bodies along with their families.

—Jennifer Battaglia, MS
University of Rochester School of
Medicine and Dentistry, Class of 2021

THE HUMANITY OF ANATOMY

Some of the best ideas in medical education at this institution come from our medical students.

The Reflector is a remarkable Medical Humanities project created by one of our medical students, Jen Battaglia. It began in response to Jen's own experience of the anatomy lab and was intended for her classmates. This book is the product of three years of effort and dedication, energy and generosity.

Before medical school, Jen matriculated in our Master's program in Medical Humanities. After completing a biochemistry degree and premed sciences, she wanted to better understand the broader humanistic and sociocultural contexts of healthcare. She understood that scientific knowledge and skills are critical in medicine, but that the practice of medicine is also deeply human and requires different knowledge and skills.

The care of patients takes place within the foundational human relationship between a patient, who is sick, and a physician, who uses the tools of science for diagnosis and treatment of the patient. These social, cultural, and relational interactions between patients, families, communities, and colleagues are the focus of medical humanities and arts education at our institution. Using materials and methods from humanities, arts, and social sciences we develop knowledge, skills, and tools that can be applied directly to the care of patients, in team-based clinical work.

At the University of Rochester, Medical Humanities is grounded in the Biopsychosocial (BPS) model of patient care. The BPS model was formulated and taught here by Rochester internist/scientist and teacher George Engel beginning in 1977 and has become part of our institutional culture. Although deeply committed to the biomedical sciences, Engel understood that "a medical model must also take into account the patient, the social, the psychological, and other contexts in which [the patient] lives. . . ." Engel's original BPS model, based on systems theory, extended beyond the dynamic interactions between the biomedical, psychological, and social to include family community, culture, nation, and biosphere. Engel focused on the patient as person and their experience of illness in the doctor/patient relationship. In our educational work, Division faculty have recentered Engel's model to include both patient and physician as persons in all of these contexts.

By looking at medicine from diverse disciplinary perspectives like history, literature, visual arts, ethics, religions, gender, race, disabilities, and cultural studies, our learning activities allow students to engage respectfully and reflectively with a diversity of perspectives, values, and identities—their own and those of others. We work collaboratively with our colleagues in clinical courses throughout the curriculum to integrate the sciences with human values and perspectives in healthcare.

The Reflector exemplifies what we do in Medical Humanities. This book is not decorative, or ancillary; it is directly relevant to the course, and fully integrated into the anatomy lab learning. Students are using this throughout the course, and they are grateful. It is an interactive educational tool that supports the learning goals of the anatomy course. It also provides a space for students to learn about anatomy through a different lens, and also to think, to feel, and to reflect on the anatomy experience personally and professionally at the beginning of their training as physicians. It has been helpful to our students, and I hope it will be helpful to other students at other institutions, too.

—Stephanie Brown Clark, MA, MD, PhD
Associate Professor, Medical Humanities
Director, Division of Medical Humanities & Bioethics Director,
MS Program in Medical Humanities
University of Rochester School of Medicine & Dentistry
stephanie_brownclark@urmc.rochester.edu
https://www.urmc.rochester.edu/medical-humanities.aspx

Block I

CHAPTER 1

Back and Spinal Cord

I. Back

Learning Objectives

- Reflect on the experience of meeting your donor.
- Notice the feeling of standing at the table with your donor.
- Think about the experience of dissecting in this first lab.
- Begin to create a productive and supportive group dynamic.

Fourteen weeks.

Fourteen weeks to dissect an entire human body. To learn thousands of new terms, essentially a language. In 14 weeks, we pried off the posterior aspect of the vertebrae and saw the beauty of the spinal cord. We dislocated the shoulder and separated the branches of the brachial plexus, tracing them down the arm. We sawed down the midline of the face, delicately revealing nerves, following them to ganglia. We used all our strength to bisect the pelvis, completing the course with body parts randomly strewn across the table. Fourteen weeks to push through, so we compartmentalize, but not too much.

This experience defines growth—academic, yes, but personal too. From the first day, timid and scared, to the days we thought there was too much to learn, too much to do. Ending confident and curious, wishing we could rewind to day one. Rewarding and humbling, mere words do not do it justice. But to Ronald: you made the ultimate sacrifice for these fourteen weeks. We promise your gift will transcend time and space. Thank you.

—Jessica Forman, Class of 2023

A. Vertebral Column

Atlas (C1): named for the Atlas of Greek mythology, who was condemned to hold up the sky for eternity. Likewise, the atlas supports the globe of the head.

Axis (C2): axle or pivot; the pivot around which the first cervical vertebra, the atlas, rotates.

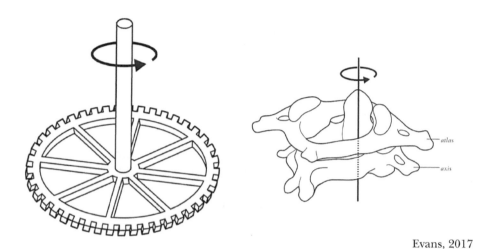

Evans, 2017

B. Back Muscles

Trapezius: Greek trapezoid = a quadrilateral with two parallel sides.

Dorsi: Latin = towards or on the back.

Bidloo, 1690

II. Deeper Back Muscles and Spinal Cord

Learning Objectives

- Think about your role in your lab group.
- Reflect on the experience of examining the back and spinal cord.
- Draw a typical spinal nerve showing its efferent (motor) and afferent (sensory) components and how it originates from the spinal cord.

Perhaps of all the remarkable surprises that I encountered during my first day as a student in the human anatomy lab, nothing was as striking as first seeing the smallness of the diameter of the spinal cord. Not much thicker than a pencil, this rope of nerve cells and fibers coursing up and down the midline of the back carries all the converging, two-way traffic of information that tells the brain what the body is experiencing and, conversely, allows the brain to command the body to respond. All those complex sensations and motor patterns carried by a delicate, white cord of tissue thinner than your little finger! Any trepidation I had moments earlier about the task before me had now given way to astonishment, humility, curiosity, and gratitude. I was hooked!

—David Kornack, PhD,
Instructor of Human Anatomy

Rhomboids: name is derived from the shape—similar to a rhombus.

A. Intermediate Muscles

Serratus: Latin = saw-toothed; describes muscles with fleshy digitations resembling the teeth of a saw; these muscles attach to multiple ribs.

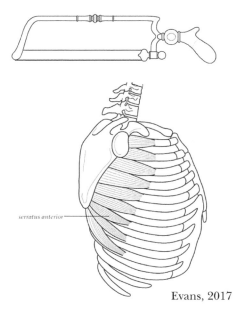

serratus anterior

Evans, 2017

B. Deep Muscles

Splenius: Greek splenion = bandage.

Capitis: derived from Latin caput = head.

Cervicis: Latin = of the neck.

Spinalis: Latin = of or belonging to the spine.

Ligamentum nuchae: derived from French nuque = nape or back of the neck.

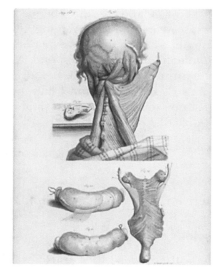

Cowper, 1698

C. Vertebral Canal and Spinal Cord

Vertebrae

Annulus: diminutive of Latin anus = finger or signet ring.

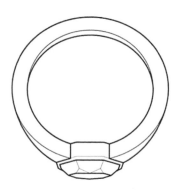 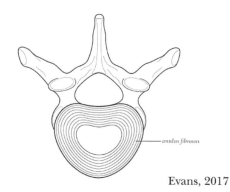

annulus fibrosus

Evans, 2017

Ligaments

Supra: Latin prefix = superior to.

Intra: Latin prefix = within.

Ligamentum flavum: Latin = yellow ligament.

Meningeal Layers

Mater: Latin = mother.

Dura mater: Latin = tough mother.

Arachnoid mater: derived from Greek arachne = spider; the suffix -oid = similar to.

Pia mater: Latin = tender mother.

Epidural: derived from the Greek prefix "epi-" = upon; hence, external to dura mater.

Denticulate: derived from Latin dens = tooth; hence, having small tooth-like projections.

Spinal Cord

Conus medullaris: Latin conus = a pinecone; medullaris = from Latin marrow (e.g. bone marrow) that was recognized to be the soft, central part of bone. When early anatomists looked at the vertebral column, the spinal cord (originally named the medulla spinalis) looked like bone marrow.

Filum terminale: derived from Latin filum = thread.

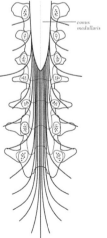

Evans, 2017

Cauda equina: Latin = the horse's tail; a bundle of nerve fibers extending from the end of the spinal cord.

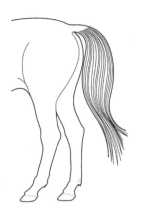 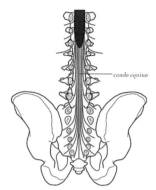

Evans, 2017

Spinal Nerve

Ramus (plural, rami): Latin = branch.

Check Your Understanding: draw a typical spinal nerve showing its efferent (motor) and afferent (sensory) components and how it originates from the spinal cord.

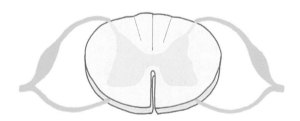

CHAPTER 2

Upper Limb

I. Shoulder

Learning Objectives

- Reflect on the experience of starting to dissect the upper extremity.
- Notice how you have been referring to your donor.
- Recognize the interactions you have had with other donors at this point.
- Reflect on your lab group's dynamic.

A. Introduction

I have always relished the tending of gardens. I seriously contemplated a career in botany.

I am fascinated by the germination of seeds. Delicate and vulnerable, seedlings are continually impacted by variables that change the course of their development. Nurture and support yields robust and healthy plants, capable of achieving their full potential. With neglect, seedlings wither. Tending my gardens, some seedlings mature, yielding bright and colorful flowers. Others mature into herbs that enchant summer meals. Some bear delicious fruits and vegetables. Often, I pause with wonder to examine features that bees, hummingbirds, butterflies, and finches find irresistible. All too soon, warm amber sun is replaced by crisp air that becomes laden with fog at each exhalation. Stems wither, leaves turn yellow, then brown, and grow brittle. I tend them one last time, gathering their remains and placing them in my compost pile. I celebrate them and prepare them to nurture future seasons' growth.

I chose to pursue a career in anatomy instead.

I tend to the donors in the peaceful solace of the anatomy labs. Rows of dissection tables await teams of students, aligned like rows of plants awaiting hovering bees. I nurture seedlings in that space, supporting the realization of their full potential. Dissections completed with perseverance, intent, and burgeoning skill are the bees, butterflies, hummingbirds, and finches that grace my gardens. They stir me to pause and wonder at the beauty and economy of design of the human form and the vigor of the learning that happens in this space. I marvel at the evolution of seedlings into mature forms, readying to share unique skills and talents with the world. I tend to the donors one last time, celebrating their gifts and preparing them for their final rest. Their work on this earth is complete, having nurtured a future seasons' growth.

—Martha J. Gdowski, PhD,
Instructor of Human Anatomy

Basilic: derived from Arabic al-basilik = inner; the basilic vein is on the medial (inner) side of the arm and forearm. This term was originally thought to be of Greek origin, derived from the Greek basilikos = royal (king-sized.)

Cephalic: derived from Arabic al-kifal = outer; the cephalic vein is on the lateral (outer) side of the arm and forearm. This term was a mistranslation of Arabic. It was originally thought to be derived from Greek kephale = head.

Cubital: derived from Latin cubare = to lie down; due to the Roman habit of reclining on the elbow even when eating.

B. Back and Shoulder Regions

Check Your Understanding: the superficial muscles of the back are associated with the upper limb. Use this space to review the muscles of the back.

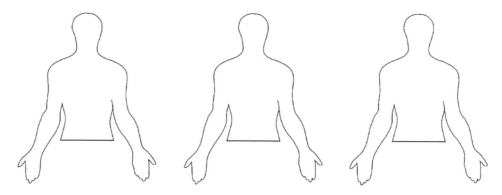

Acromion: highest point of the shoulder; from akron = summit, or peak, and omos = shoulder.

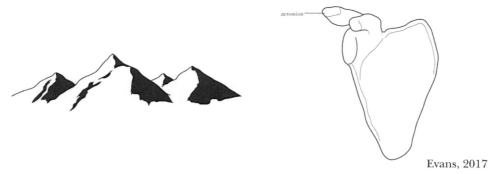

Evans, 2017

Coracoid: raven-like; from korax = crow or raven; coracoid process of the scapula resembles a raven's beak and provides attachment for a number of muscles.

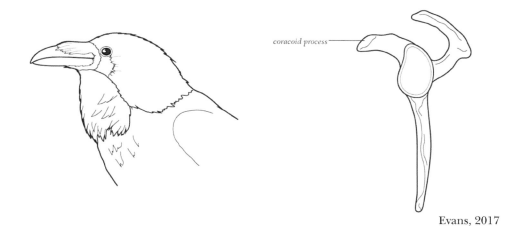

coracoid process

Evans, 2017

Circumflex: Latin circum = around, and flexere = to bend; hence, bent or bend around.

Teres: Latin = rounded, cylindrical.

Deltoid: uppercase delta (shown below) is the 4th letter of the Greek alphabet; describes the inverted shape of the deltoid muscle.

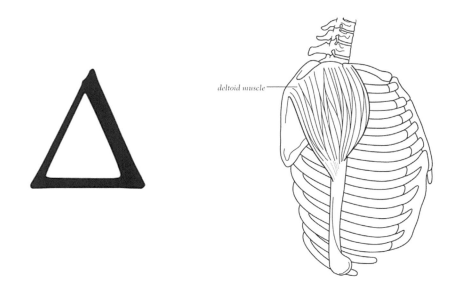

deltoid muscle

Evans, 2017

Triceps: three-headed; from tri = three, and caput = head.

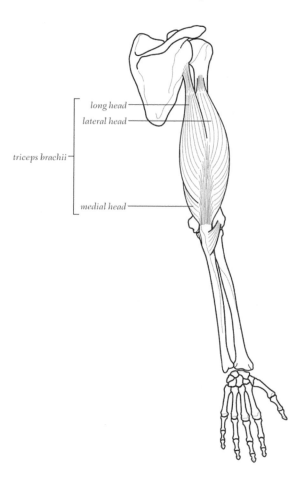

Evans, 2017

Check Your Understanding: use this space to label the bony landmarks of the scapula, and the muscles of the rotator cuff.

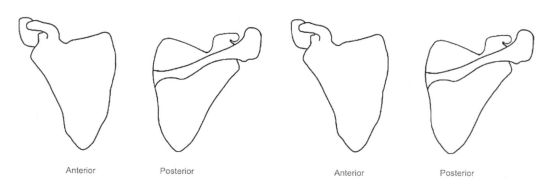

Anterior   Posterior   Anterior   Posterior

II. Axilla and Arm

Learning Objectives

- Think about how it feels to dissect when your donor is supine, instead of prone.
- Draw the contents of the axillary sheath and the brachial plexus.
- Draw the muscles and blood supply of the anterior and posterior compartments of the arm.

Dissecting a body is weird. Some days I am simply astounded—the intricacies of the arterial branches supplying the trunk, the brachial plexus crossing and weaving into that landmark M, the details within the chambers of the heart. Such purpose contained in something so small. I want to be closer, taking it all in. Most days, I am ready to complete the task. Cut here, rotate this, find X. But there are moments when that is different. Moments when I see the dry, cracked lips of my donor's face and wonder if he was as thirsty and uncomfortable as my grandpap while he was sick. Moments when we find something—a variation in artery placement, a tiny growth on her brain—and know something about her that she never knew about herself. Seeing her massive heart, her defibrillator, her restructured vasculature—and wondering about the likely pain and fear—gives rise to insights into her life revealed solely by her organs. Feeling protective of her prolapsed uterus while my classmates crowded around. Appreciating the irony between restoring her shoulder to a more comfortable, more natural position while being the most excited to use the bone saw. What a strange and beautiful task at hand.

—Sara Peterson, Class of 2022

A. Anterior Wall

Pectoralis: from Latin pectoris = front of the chest.

B. Axilla and Axillary Contents

Plexus: from 'to plait' = to weave strands of material (hair, straw, or flax) into a braid, cord, or rope; a network of interconnecting nerves or vessels.

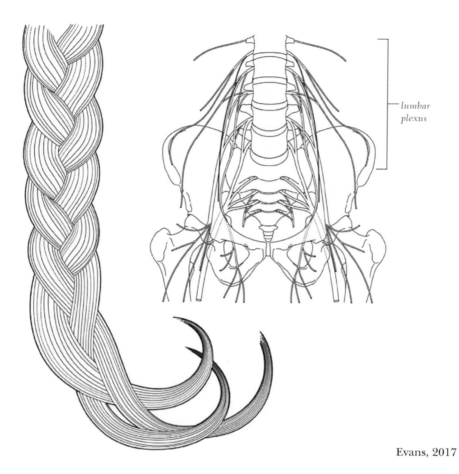

lumbar plexus

Evans, 2017

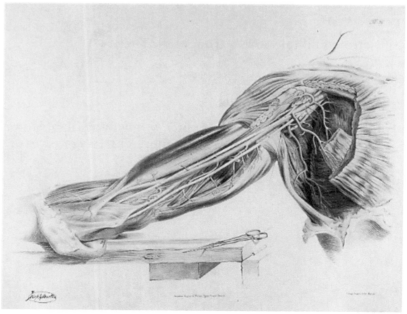

Quain, 1844

Check Your Understanding: use the diagrams below to draw the axillary sheath and its contents. Also, identify the borders of the axilla.

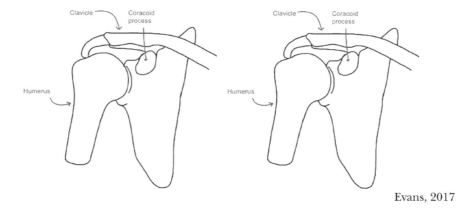

Evans, 2017

Check Your Understanding: use this diagram to draw the brachial plexus, from the roots, to the five terminal branches.

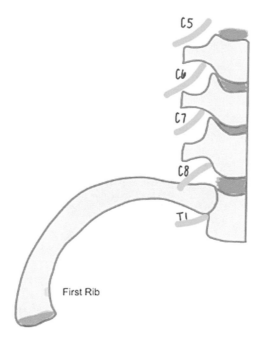

C. Arm

Biceps: two-headed; from bi = two, and caput = head.

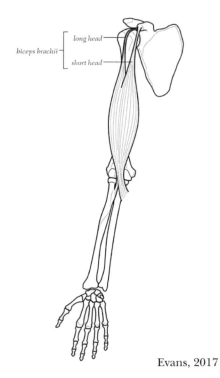

Evans, 2017

Check Your Understanding: use the diagrams below to label the muscles, blood supplies, etc., of the anterior compartment of the arm.

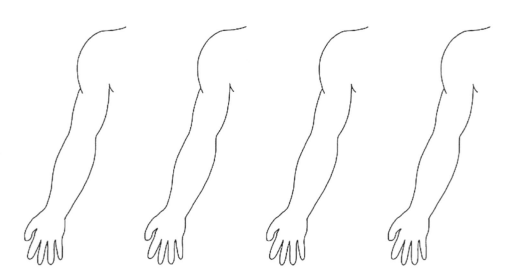

Check Your Understanding: use the diagrams below to label the muscles, blood supplies, etc., of the posterior compartment of the arm.

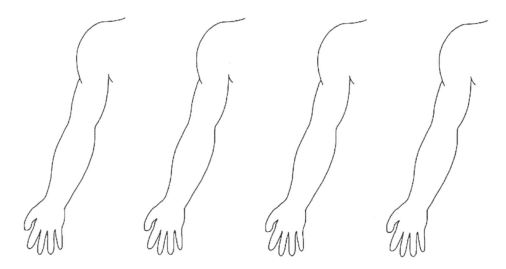

III. Flexor Forearm

Learning Objectives

- Reflect on the experience of learning your donor's name.
- Notice how your lab team supports one another and notice the role that you play in this team.

Anatomy Lab, 1984

Stepping into the room filled with tables of sheet-covered forms
We approach tentatively
We look around at one other for support

Some say little, others joke awkwardly
The covers are pulled back by each team
I can't remember...
Was our body a woman, a man?

No image I can conjure
What I remember is the smell,
The pervasiveness of the formalin
That clings to every crevasse of our being all year
And the numbing of emotional response

There are a few nods to the fact that these were people
The whispered queries about who they might be
Where they come from
How they got here

Years later, during a discussion of that time with a colleague
He shared that surgery, as a previously prized profession
Was lost to him that year, when
The family of the cadaver they were dissecting turned up mid-class to claim him

What I remember is the first cut
The inflicting of a wound on this still body
Now I remember—a woman she was
The abdomen both smooth and wrinkled as we made that first cut

"Look", my classmate muttered under his breath
"underneath the skin we all look the same"
Something in the tone did not signal this as
A comment recognising our common humanity

I pulled myself closer in
This dissecting of black bodies
In Apartheid South Africa
Had never felt more violating
      —Patricia Luck, MD, Division of Medical Humanities and Bioethics

A. Bony Landmarks

Check Your Understanding: use this diagram to label the landmarks on the humerus, radius, and ulna.

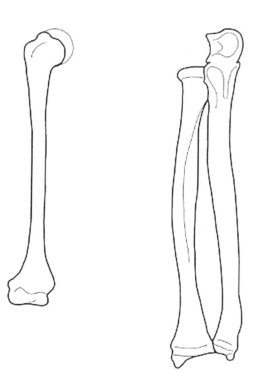

Humerus: Latin = the arm-bone.

Epicondyle: Greek epi = upon, and kondylos = knuckle; prominence on the part of the humerus that looks like a knuckle.

Capitulum: derived from Latin caput = head.

Trochlea: Greek trochilia = a pulley.

Olecranon: Greek olene = ulna, and kranion = upper part of head; upper part of the ulna.

Tuberosity: Latin tuber = a swelling or lump, usually large and rough.

Styloid: Greek stylos = an instrument for writing, and eidos = shape or form; hence a pen or pencil-like structure.

B. Forearm Structures

Carpi: Greek = wrist.

Digitorum: Latin digitus = a finger or toe.

Palmaris: Latin = palm.

Pollicis: possessive derivation of Latin pollex = thumb; hence, of the thumb.

IV. Palm

Learning Objectives

- Reflect on dissecting your donor's hands.
- Draw the tendons and nerves that pass through the carpal tunnel.
- Draw the muscles that contribute to the thenar and hypothenar eminences.
- Draw the motor and sensory innervation of the hand, and the muscles and tendons of the palm.

A. Introduction

I had three hands.
My right was holding the blade.
The left was pinching fancy tweezers.
The third was being taken apart.

I had to focus on all three hands at once.
Where the blade bit, where the pincers pulled.
Where the tendons took their path.
Using my hands to understand the inside of another.

There was a list to get through
muscles
nerves
compartments
There was no time to notice
the atrophied thenar eminence
the crooked fingers
the small tattoo on the wrist
the wrinkled skin

There was no time to see
the toll of decades of arthritis
the other hands this one might have held
the history written on her skin
the things these hands may have done

the third hand spoke its own story,
if the other hands could hear it.

Who was to say that this story was any less
Important than the items on the list.

—Antoinette Esce, Class of 2019

B. Bony Landmarks

Pisiform: Latin = pea-shaped.

Hamate: from Latin hammus = hook.

Triquetrum: Latin triquetrus = three-cornered.

Lunate: Latin lunatus = crescent, or halfmoon-shaped.

Scaphoid: Greek skaphoeides = boat-shaped, hollow.

Phalanges: from Latin phalanx = row of soldiers.

Check Your Understanding: use this diagram to draw the eight carpal bones.

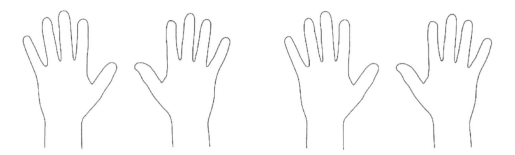

C. Wrist

Thenar: Greek = palm of the hand.

Check Your Understanding: use these diagrams to draw the flexor retinaculum and the contents of the carpal tunnel.

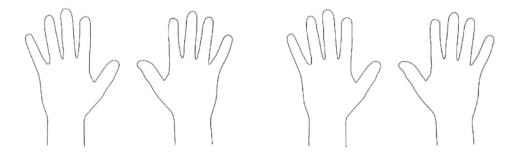

D. Digits

Lumbrical: Latin lumbricus = worm; hence, worm-shaped muscles.

Digiti: Latin digitus = a finger or toe.

Check Your Understanding: use these diagrams to draw the innervation and blood supply of the palm.

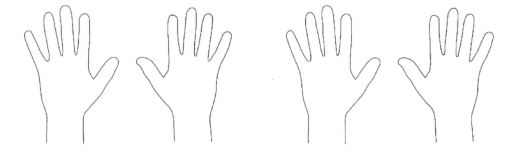

V. Extensor Forearm and Dorsal Hand

Learning Objectives

- Draw the tendons that form the anatomical snuff box, the extensors of the forearm, and the innervation and blood supply of the forearm.

Pieces of Skin

I am connected with the soul that left
But she will never know her new cuts

As I slice and dice through reflections bisected connections left intact
I pause and remember how little I know

I try my best to imagine what she carried with her hands
as I cut pieces off of both of us that day
Her new wounds made entirely for my benefit
And I still remember so little of what I once knew

I am taught that every emotional pitfall is for my career but I can't help
but pray for the remnants I washed away
For me to make myself the subject of these cuts is selfish
I am connected with the soul that left

—Zain Talukdar, Class of 2023

A. Anatomical "Snuff Box"

The name anatomical "snuff box" comes from using the depression on the dorsal hand as a means of placement for the inhalation of powdered tobacco (dry snuff) and was first described in the medical literature in 1850 (S. Hallet & J.V. Ashurst)

Check Your Understanding: draw the tendons that form the anatomical "snuff box."

B. Superficial and Deep Extensors

Indicis: Latin index = a pointer.

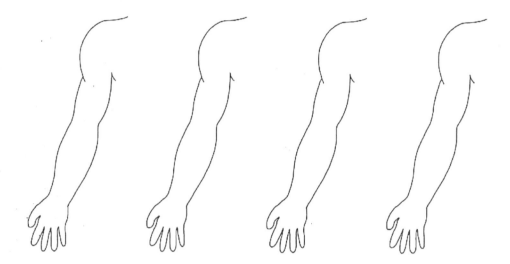

Check Your Understanding: draw the superficial and deep extensors of the forearm.

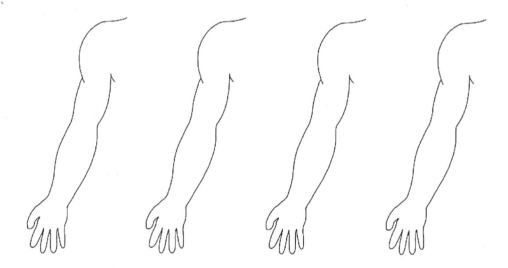

VI. Joints of the Upper Limb

Learning Objectives

- Prepare with your team to see your donor's face.
- Take a step back to observe your donor's signs of aging.

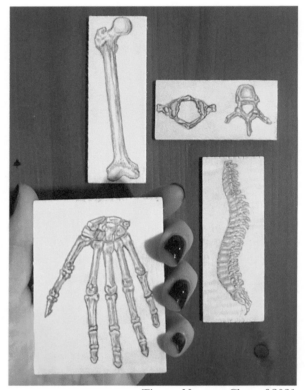

Tianna Negron, Class of 2021

A. Shoulder Joint

Glenoid: Greek glene = socket, eidos = shape or form.

Labrum: Latin = rim.

B. Elbow Joint

Collateral: Latin con = together, and latus = side, hence, alongside.

Block I

BOLD TERMS

I. Back and Spinal Cord

Scapula
Spine
Acromion
Superior angle
Inferior angle
Medial/vertebral border
Iliac crest
Posterior superior iliac spine
Occipital bone
External occipital protuberance
Nuchal lines
Cervical vertebrae
Transverse process
Transverse foramen
Spinous processes
Atlas
Axis
Vertebral foramen
Vertebral canal
Thoracic and lumbar vertebrae
Body
Vertebral arch
Pedicles
Laminae
Facets
Superior articular process
Inferior articular process
Spinous process
Rib and its head
Rib tubercle
Fibrocartilaginous intervertebral discs
Intervertebral foramen
Neurovascular bundles

Accessory nerve
Trapezius
Latissimus dorsi

II. Deeper Back Muscles and Spinal Cord

Rhomboids (major and minor)
Levator scapulae
Serratus posterior superior
Serratus posterior inferior
Splenius capitis
Splenius cervicis
Ligamentum nuchae
Semispinalis capitis
Erector spinae
Iliocostalis
Longissimus
Spinalis
Transversospinalis muscles
Interspinous ligaments
Ligamenta flava
Epidural space
Vertebral venous plexus
Dura mater
Arachnoid mater
Subarachnoid space
Pia mater
Denticulate ligaments
Anterior and posterior roots and rootlets
Dorsal root ganglion (spinal ganglion)
Conus medullaris
Cauda equina
Filum terminale
Dermatome

III. Shoulder

Basilic vein
Cephalic vein
Median cubital vein
Scapula
- Acromion
- Spine
- Supraspinous fossa
- Infraspinous fossa

- Glenoid cavity
- Supraglenoid tubercle
- Infraglenoid tubercle
- Coracoid process
- Scapular notch

Humerus
- Head
- Greater tubercle
- Lesser tubercle
- Intertubercular sulcus (bicipital groove)
- Deltoid tuberosity
- Sulcus for radial nerve (spiral groove)

Deltoid muscle
Axillary nerve
Posterior circumflex humeral artery
Quadrangular space
Triceps brachii
- Long head
- Lateral head
- Medial head

Teres minor
Teres major
Radial nerve
Deep brachial (profunda brachii) artery
Supraspinatus
Infraspinatus
Suprascapular nerve
Suprascapular artery
Superior transverse scapular ligament
Rotator cuff

IV. Axilla and Arm

Basilic and cephalic vein
Median cubital vein
Pectoralis major
Deltopectoral triangle
Lateral pectoral nerve
Medial pectoral nerve
Pectoralis major and minor
Thoraco-acromial artery
Serratus anterior muscle
Biceps brachii
- Long head
- Short head

Coracobrachialis
Axillary artery
Axillary veins
Superior thoracic artery
Lateral thoracic artery
Subscapular artery
Thoracodorsal artery
Circumflex scapular artery
Posterior circumflex humeral artery
Anterior circumflex humeral artery
Brachial plexus
- Lateral cord
- Medial cord
- Posterior cord
Musculocutaneous nerve
Median nerve
Ulnar nerve
Axillary nerve
Radial nerve
Medial cutaneous nerve of the forearm
Lateral cutaneous nerve of the forearm
Subscapular nerves
Thoracodorsal nerve
Long thoracic nerve
Brachialis
Radial artery
Ulnar artery

V. Flexor Forearm

Humerus
- Medial epicondyle
- Lateral epicondyle
- Capitulum
- Trochlea
Suprascapular nerve
Suprascapular artery
Superior transverse scapular ligament
Rotator cuff
- Olecranon fossa
Radius
- Head
- Neck
- Tuberosity
- Styloid process

- • Interosseous border

Ulna

- • Head
- • Olecranon
- • Interosseous border

Pronator teres
Flexor carpi radialis
Palmaris longus
Flexor digitorum superficialis
Flexor carpi ulnaris
Cubital region
Radial artery
Ulnar artery and nerve
Brachial artery
Median nerve
Common interosseous branch of ulnar artery
Anterior interosseous branch
Interosseous membrane
Brachioradialis
Superficial branch of radial nerve
Deep branch of radial nerve
Supinator muscle
Flexor digitorum profundus
Flexor pollicis longus
Pronator quadratus

VI. Palm

Thenar eminence
Hypothenar eminence
Scaphoid
Lunate
Triquetrum
Pisiform
Trapezium
Trapezoid
Capitate
Hamate
Metacarpal bones
Phalanges
Proximal phalanx
Middle phalanx
Distal phalanx
Palmar aponeurosis
Palmaris brevis

Superficial palmar arterial arch
Ulnar artery
Common digital arteries
Proper digital arteries
Flexor retinaculum (transverse carpal ligament)
Carpal tunnel
Median nerve
Recurrent branch of the median nerve
Abductor pollicis brevis
Opponens pollicis
Flexor pollicis brevis
Abductor digiti minimi
Opponens digiti minimi
Flexor digiti minimi brevis
Lumbrical muscles
Extensor expansion
Pronator quadratus muscle
Deep branch of the ulnar nerve
Adductor pollicis muscle
Deep palmar arterial arch
Palmar interossei muscles

VII. Extensor Forearm and Dorsal Hand

Abductor pollicis longus
Extensor pollicis brevis
Extensor pollicis longus
First dorsal interosseous muscle
Extensor retinaculum
Brachioradialis
Extensor carpi radialis longus
Extensor carpi radialis brevis
Extensor digitorum
Extensor digiti minimi
Extensor carpi ulnaris
Abductor pollicis longus
Extensor pollicis brevis
Extensor pollicis longus
Supinator
Extensor indicis
Deep branch of the radial nerve
Dorsal interosseous muscles
Extensor expansion
Lumbricals
Interossei

VIII. Joints of the Upper Limb

Subscapularis muscle
Glenohumeral ligaments
Glenoid cavity
Glenoid labrum
Tendon of the long head of the biceps
Coracoacromial ligament
Coracoclavicular ligament
Ulnar collateral ligament
Radial collateral ligament
Anular ligament

Block I

REFLECTIONS

Chapter 1

I. Back

What did it feel like to be in the anatomy lab for the first time?
Looking forward, what are some challenges that you can imagine you will face in the lab?
How will you address these challenges?
Aside from the anatomy that you see in the historical images shown above, what else do you see?

II. Deeper Back Muscles and Spinal Cord

In what ways does your group function as a unit?
What are some strengths that you can bring to your lab group?

Chapter 2

I. Shoulder

How do you refer to your donor? Do you use donor, cadaver, or another name, and why?
What has been your experience interacting with other donors?
In what ways have you made positive contributions to your lab team?
Are these the ways that you had expected?

II. Axilla and Arm

Was it different to dissect with the donor in a supine, instead of a prone, position?
What are your donor's muscles like? What colors do you see? Have you noticed any scars, fractures, or calcifications?
What do your observations tell you about your donor?
What do you think it was like to live in your donor's body?

III. Flexor Forearm

How did it feel to have a palmaris longus or not?
What was it like to learn your donor's name?
What was it like to learn your donor's profession?
What was it like to learn your donor's age and cause of death?
Did you and your lab group members have similar or different reactions to learning information about your donor?

IV. Palm

What was it like for you to dissect the hand?

V. Extensor Forearm and Dorsal Hand

Is it hard to imagine your donor as a live body?
Based on your observations, what would it be like to live in your donor's body?
How do your donor's hands compare to the hands of other donors that you have studied with?

VI. Joints of the Upper Limb

What signs of aging do you see on your donor?
What signs of aging do you see on other donors?
What was it like to see your donor's face? How did you feel before the face was revealed? Was it what you expected?
Have you seen the faces of other donors? If so, how do they compare to your donor's face?
How did your group members support one another during today's lab?

Post-Exam Reflection

What was the process of taking the exam like for you?
What was it like to stand beside other donors during the exam?

Block II

Chapter 3

Thorax and Root of the Neck

I. Thoracic Wall

Learning Objectives

- Reflect on the donors' decisions to donate their bodies.
- Think about how your views of the anatomy lab have changed or remained the same since the first day of lab.
- Consider the reflection questions with your lab group. Listen to each other's perspectives and learn from one another.

A. Introduction

Who Are You

Who are you
I say "who are you"
Not "who were you"
Because
You are still you
Different yes
Transcendent
Yet still so near

Why is that I feel
I should kneel
Whisper a prayer
Offer a blessing
Burn incense
In your presence
Honoring this sacrament
This gift
Of vulnerability
Contained
In your body
Bringing forth holiness
Out of death and sorrow

Yet it is so hard
For I am rushed

When visiting
The great houses of worship
Or the intimate recesses
Of a person's home
Does one rush through
Only noting
What is useful
To one's self?

Who are you
Who have invited us
To know you
As no one else has
Not even your beloveds

Who are you
Who allowed us
To disrobe you
Of that fleshy vestment
You presumed
Totally, utterly yours

Who are you
That by your graceful consent
Transformed
What we do
From violation and transgression
Into communion and growth

Who are you
Whose hands
Now still adorned with pink nail polish
Once held a child
Dug in the earth planting seeds of hope
Spread fingers wide in the wind
Clenched in furious rage
Are now are so tight
That i must cut your tendons
To release the tension

Who are you
Whose cheeks
Wrinkled with smiles
Received countless kisses
Bathed in tears of loss
Burnt red by the sun forgotten in revelry

Yet lie severed
For there are structures
Beneath

Who are you
Whose heart
Quickened in the fear and excitement
Of not knowing
Stilled as only hearts can, to listen
Fractured in countless moments of loss
Healed in ways unimaginable, indescribable
Offered and received
The one thing
We all desire
Love
And now it lays on a table
Pin holding back its flap
A tag asking: "what valve is this"

Who are you
Whose eyes
Beheld beauty and ugliness
Searched in longing
Closed to see beyond the physical
Blurred with mingled tears of joy and sorrow
Rose up in hope
Now here one eye lies
Sliced in two
And I am disappointed
Not to find within them
The path to your soul

Who are you
Whose mouth
Broke with laughter
Smiled seductively
Hung open astonishedly
Thinned with anger
Keeping back violent words
Pursed in disappointment
Tightened in enduring pain
Whispered trusted secrets
Yet now your lips
Are dry
And dark

Who are you
Whose mind
Grappled

With paradoxes
Sat with mysteries
Tried to give form
To the ineffable
Which your head
Now embodies
Being cleaved in two
Whispering the truth
That the human spirit
Is even more nuanced
Multifaceted in its identities
And secrets
Than I could have imagined

Who are you
That by the gift
Of your body
Still tinged with your
Being
Teach me
Not just about
Flesh and bone and viscera
But also about
Some deeper quality
Of being human
That intermingling
The transmutation
That rhythmic dance
Of light and dark

What shall I call you
Guide?
Priestess?
Patroness?
Accompanier?
Teacher?
Seeker?
Prophet?
Are you the Virgil
To my Dante?
Guiding me to
The depths
To
Places I would rather
Not go
Questions I would rather
Not ask
Realities I would rather
Not face

Yet which are so vital
To truly being human
To connecting at painful places
To offering true healing
Even when a cure
Is not possible
Standing in the breach
Against incessant, inevitable
Waves of suffering
And death

For how much
If that type of suffering
Is living?
Who are you
You who teach me
Not with words
But with something
Far more eloquent
Your being
Your very body here
Dead
Yet still a gift
Drawing forth
Life
Growth
Communion
From the darkness
By the work of your spirit
Of your vulnerability
Of your trust in me

What greater paradox
What greater lesson
What greater question
What greater mystery
Is there to plumb?

Like any good teacher
You ask a question
But don't give me the answer
Rather
Reminding me
By your presence
To stay with the question
Even when it is hard
Taking breaks
Not to run and hide
But to rejuvenate

So as to engage the question
Anew
Not to answer
And box it in
With comfortable certitude
But to explore
With openness and curiously
This mingling of
Life and death
Joy and sorrow
Love and suffering
Mystery and certainty

Feeling the pained,
Hopeful beat
Of my heart next to yours
You ask me
Who are you?

—Paul Kim, Class of 2024

Kyphosis: Greek kuphosis = bent, or hunchbacked.

Lordosis: Greek lordos = bent backwards.

Scoliosis: Greek skolios = bent.

B. Bony Landmarks

Sternum: Greek sternon = chest.

Clavicle: Latin clavicula = a small key.

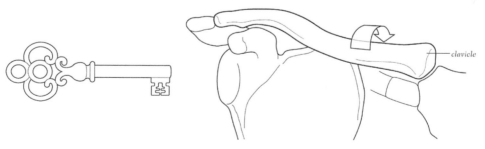

Evans, 2017

Manubrium: from Latin = haft, or handle.

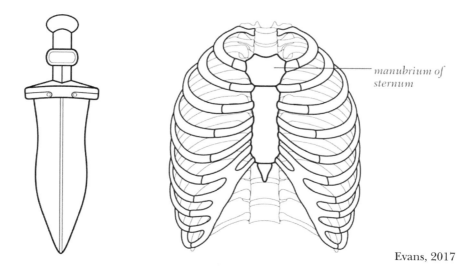

manubrium of
sternum

Evans, 2017

Xiphoid: from Greek xiphos = sword, Greek suffix -oid = similar to, form, resemblance, shape, likeness; hence, sword-shaped.

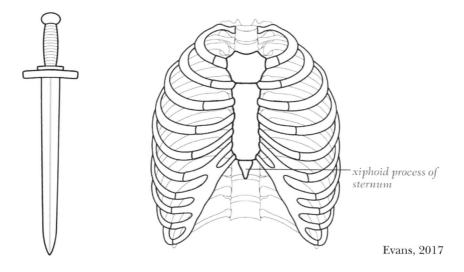

xiphoid process of
sternum

Evans, 2017

C. Female Breast

Areola: a small circular courtyard or open space of ground in front of a Roman house.

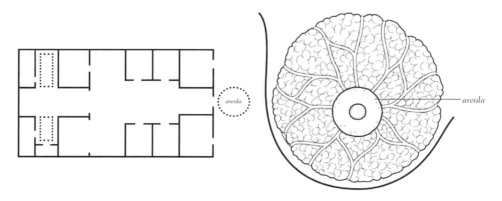

Evans, 2017

D. Muscles, Nerves, and Vessels

Costal: Latin costa = rib.

II. Pleural Cavities and the Lungs

Learning Objectives

- Draw the features of the lungs and the hila.
- Imagine what it will be like to auscultate your patient's lungs.
- Compare and contrast the lungs of different donors.

A. Introduction

Visceral: Latin viscus = internal organ.

Parietal: Latin parietalis, pertaining to paries = wall.

Diaphragm: Greek dia = across, and phragma = a wall or fence.

B. Pleural Cavities

Pleura: Greek = side of the body, rib.

C. The Lungs

Phrenic: from Greek phren = diaphragm, mind (the mind was once thought to lie in the diaphragm).

Fissure: from Latin findere = to split.

Hilum: from Latin = little thing, trifle; the scar left on a seed coat by its attachment to the plant.

Alveolus: Latin = a small cavity.

Check Your Understanding: draw the distinguishing features of the right and left lungs and their respective hila.

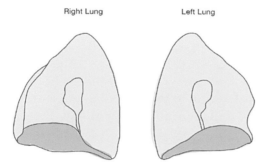

Right Lung   Left Lung

III. Mediastinum and Heart

Learning Objectives

- Think about holding different human hearts in your hands.
- Reflect on the complexity of the heart and what it will be like to learn how to listen to heart sounds.

A. Introduction

Today I held your heart. I put my fingers
around your vessels. I washed until
they glowed and your blood shook out
in so many shades of rust. And, yes, it's true,
only the other morning I broke
your spine. I shivered at your bony ridges,
the color of so many whitened trees
in winter. Afterwards, I carved into
your wrinkles until I found
that startled dark pink, and I uncurled
your stiff fingers to lay my thumb
on your palm, your tendons drawn

under the weak October light.
I want you to know that this is beautiful—
your barrel chest and wasted thighs,
your singing neck and painted nails,
even the crusts on your skin and the hair
on your upper lip. I want you to know
that of those who have held you close,
I have held you closer, my hands
cradled around your brain or pressed
warm against your ribs. In the end,
I want you to know how we smell you
on our skins as we walk to the locker room,
how we undress, our backs turned
in modesty, covering our secrets—
what we are naked and on the inside—
your body reflected in all of ours,
no perfect mirror but enough to make us
nervous, so awed and almost fearful
at the quiet pulse within us.

—Jennifer Hu, Class of 2018

Mediastinum: from Latin medius = middle, and stans = standing; hence, a median vertical partition, adjective - mediastinal.

B. Pericardium

Pericardium: Greek peri = around, and kardia = heart; hence the membranes enclosing the heart.

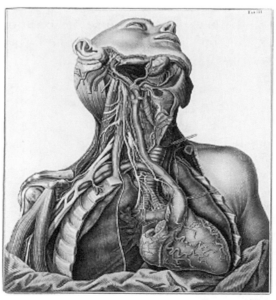

Scarpa, 1794

Serous: from French sereux, or Latin serosus = serum.

Sinus: Latin = recess, hollow space.

C. Heart and Great Vessels

Atrium: Latin = a formal hall or court, the focal point of a Roman house; a central room.

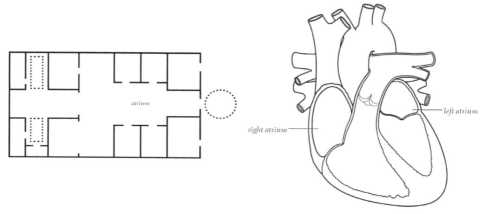

Evans, 2017

Ventricle: from Latin ventriculus, diminutive of venter = belly.

Vena cava: Latin vena = vein, cava = hollow; hence, hollow vein.

Vagus: Latin = wandering.

D. Cardiac Vessels

Corona: Latin = wreath, crown.

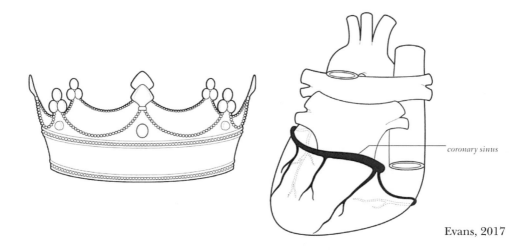

coronary sinus

Evans, 2017

E. Interior of Heart

Pectinate: Latin pectinatus = resembling a comb.

Trabeculae carneae: diminutive of Latin trabs = a beam; Latin carnea = fleshy. Hence little cords of flesh that support a structure.

Papillary: from Latin papula = small protuberance.

Moderator band (septomarginal trabecula): named because it was thought to prevent overdistension of the ventricle. It was first described by Leonardo Da Vinci in his exploration of the human body.

Commissure: Latin con = together, and missum = sent; hence, fibers that cross between symmetrical parts.

Infundibulum: Latin = funnel.

Chordae: from Latin cord = catgut, or a string from a musical instrument like a lyre (a stringed instrument like a small U-shaped harp with strings fixed to a crossbar, used especially in ancient Greece).

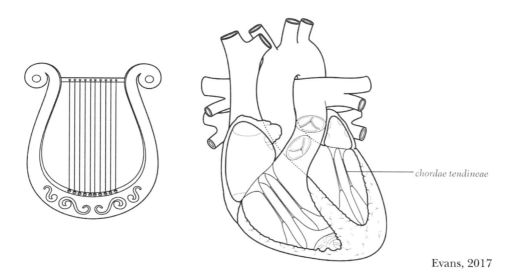

chordae tendineae

Evans, 2017

Septum: Latin saeptum = fenced in; hence, a dividing fence or partition.

Semilunar: Latin semi = half, and luna = moon; hence, having a half-moon shape.

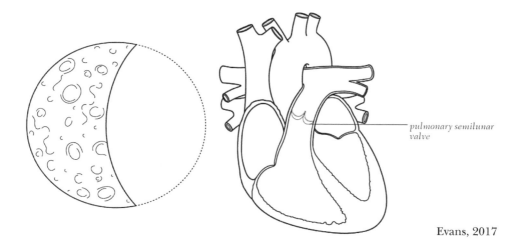

pulmonary semilunar valve

Evans, 2017

Mitral: shape of a mitre, an ornate ceremonial headdress worn by Christian bishops.

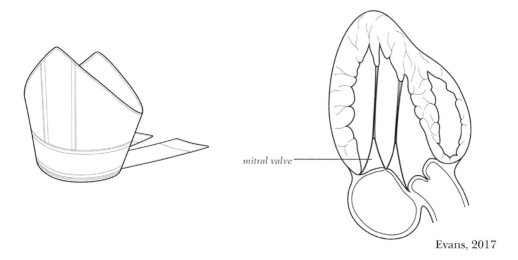

Evans, 2017

Check Your Understanding: draw the features of the heart and its blood supply on this frontal section of the heart.

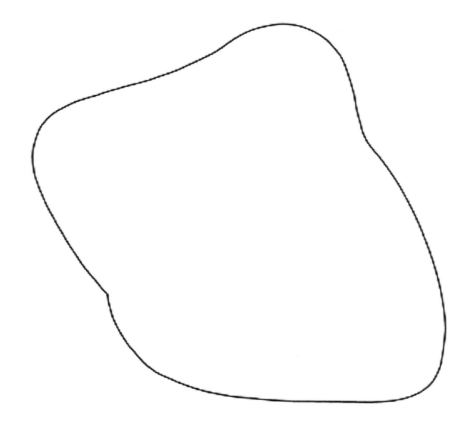

IV. Posterior and Superior Mediastinum

Learning Objectives

- Draw the great vessels in the superior mediastinum.
- Observe the empty chest cavity.
- Think about this history of the medical profession and what being a doctor means to you.

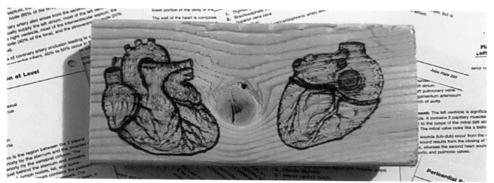

Tianna Negron, Class of 2021

A. Posterior Mediastinum

Laryngeal: relating to the larynx; Greek larynx = voice-box.

Azygos: Greek a = negative, and zygos = paired; hence, unpaired.

B. Superior Mediastinum

Brachiocephalic: Latin brachium = arm, Greek kephale = head; hence a blood vessel related to the upper limb and head.

Thymus: Greek thymos = warty outgrowth.

Trachea: Greek trakheia = rough, referring to its corrugations.

Carina: from Latin keel; a flat blade sticking down into the water from a sailboat's bottom.

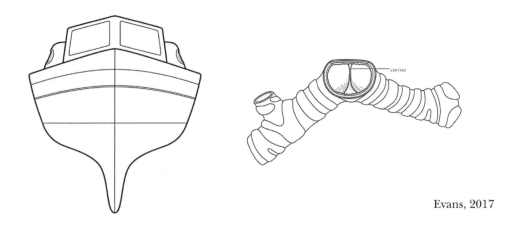

Evans, 2017

Check Your Understanding: draw the branches of the aortic arch.

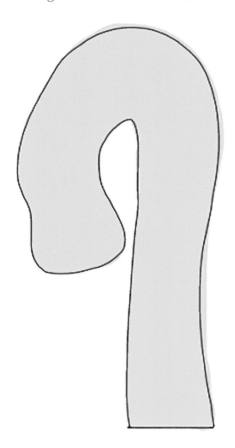

V. Anterior Triangle and Root of the Neck

Learning Objectives

- Think about the dynamic of your lab team and the role that you play in your group.

A. Introduction

Having prior experience dissecting a donor, I had already gone through the gore and disgust that naturally I would have felt. However, the experience had produced a greater sense of appreciation for the donor. The specific, delicate, and long dissections utterly surprised me and for an individual to give their body so that I may learn is a privilege that I will remember for the rest of my life. My ability to take care of patients started with the donor and the people that I help will forever be a result of the lessons I have learned through my donor. For that I am tremendously grateful.

—Filip Korityssiky, Class of 2022

Fascia: Latin = band, door frame; hence the fibrous wrapping of muscles and other structures.

B. Superficial Structures

Platysma: Greek = flat piece, plate.

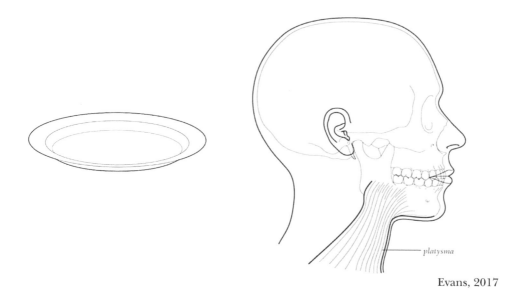

Evans, 2017

Hyoid: Greek = U-shaped; lowercase upsilon is the 20th letter of the Greek alphabet.

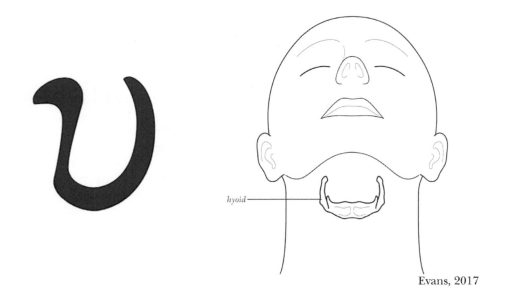

Evans, 2017

Cricoid: ring-like; Greek krikos = a ring, the suffix -oid = similar to.

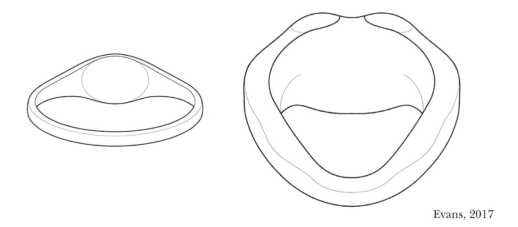

Evans, 2017

C. Muscular Triangle

Omohyoid: Greek omos = shoulder; hence, a muscle attached to the scapula and hyoid.

D. Carotid Triangle

Hypoglossal: Greek hypo = under, and glossa = tongue.

Ansa: Latin = a handle or loop.

Carotid: from Greek karoun = stupefy (the compression of these arteries was thought to cause stupor).

Jugular: Latin jugulum = neck.

Lingual: Latin lingua = tongue.

E. Submandibular Triangle

Digastric: Latin di- = twice, gaster = belly; hence, having two bellies.

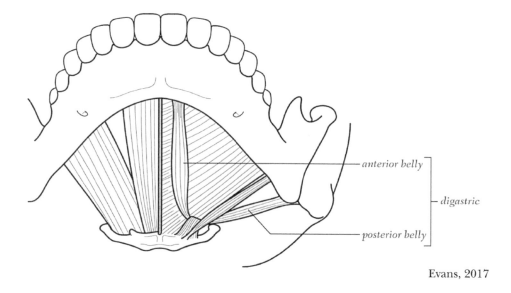

Evans, 2017

Mylohyoid: Greek mylo = molar, and hyoeides = U-shaped.

F. Thyroid Gland

Thyroid: from Greek thureoeides = shield shaped.

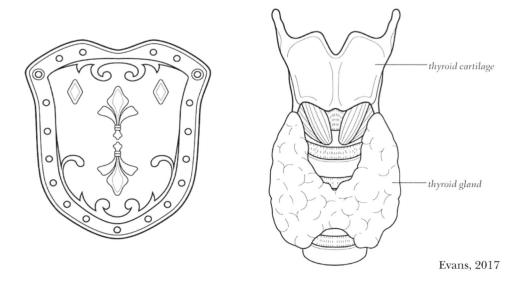

Evans, 2017

Isthmus: Greek isthmos = a narrow strip of land connecting two larger areas of land.

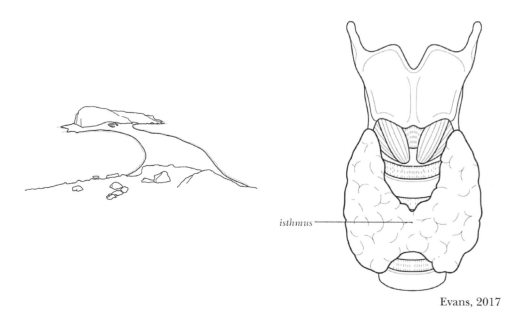

Evans, 2017

G. Parathyroid Glands

Parathyroid: Greek para = beside, and thyroid; hence, beside the thyroid.

H. Base of the Neck

Sympathetic: Greek syn = together and pathos = feeling.

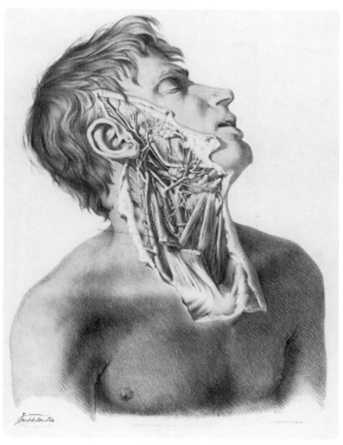

Quain, 1844

VI. Posterior Triangle of the Neck

Learning Objectives

- Think about your donor's muscles.
- Compare and contrast your donor to other donors that you have studied with.

Splenius: from Greek splenion = bandage.

Levator: Latin = a person who lifts.

Scalenus: from Latin scalenus musculus = unequal muscle.

Block II

BOLD TERMS

I. Thoracic Wall

Kyphosis
Scoliosis
Sternum
- Manubrium
- Body
- Xiphoid process

Ribs
- Head
- Neck
- Tubercle
- Body

Neurovascular bundle
Thoracic vertebra
- Body
- Pedicles
- Laminae
- Spinous process
- Articular process

Jugular notch
Sternal angle
Clavicle
Acromion of the scapula
Female breast
Suspensory ligaments
Nipple
Areola
Lactiferous ducts
Intercostal muscles
- External intercostal
- Internal intercostal
- Innermost intercostal

Internal thoracic (mammary) arteries and veins
Transversus thoracis muscle

Sternocostal joints
Intercostal nerve and vessels
Thymus gland
Right and left brachiocephalic veins
Superior vena cava (SVC)
Azygos vein

II. Pleural Cavities and the Lung

Pleural sacs
Visceral pleura
Parietal pleura
- Costal
- Mediastinal
- Diaphragmatic
- Cervical

Phrenic nerves
Abdominal diaphragm
Inferior lobe of lung
Superior lobe of lung
Oblique fissure
Horizontal fissure
Middle lobe
Bronchus
Pulmonary artery
Pulmonary veins
Hilum
Pulmonary ligament
Pulmonary lymph nodes
Lobar (secondary) bronchi
Segmental (tertiary) bronchi
Bronchopulmonary segment
Descending thoracic aorta
Intercostal arteries

III. Mediastinum and Heart

Mediastinum
Fibrous pericardium
Serous pericardium
Parietal serous layer
Visceral serous layer
Transverse pericardial sinus
Oblique pericardial sinus
Right atrium

Right ventricle
Superior vena cava
Ascending aorta
Pulmonary trunk
Left vagus nerve
Aortic arch
Recurrent laryngeal nerve
Ligamentum arteriosum
Inferior vena cava
Apex
Coronary/atrioventricular groove
Interventricular grooves
Aortic valve
Pulmonary trunk
Pulmonic valve
Right coronary artery
Marginal branch
Posterior interventricular branch (posterior descending)
Right atrial branch
Left coronary artery
Anterior interventricular (left anterior descending/LAD)
Circumflex branch
Left marginal branch
Cardiac veins
Great cardiac (anterior interventricular) vein
Middle cardiac (posterior interventricular) vein
Pectinate muscles
Crista terminalis
Fossa ovalis
Coronary sinus
Valve of the coronary sinus
Right atrioventricular (tricuspid) valve
Commissures of tricuspid valve
Chordae tendineae
Papillary muscles
Septomarginal trabecula (moderator band)
Trabeculae carneae
Conus arteriosus/infundibulum
Pulmonary semilunar valve
Left atrium
Four pulmonary veins
Left atrioventricular (bicuspid/mitral) valve
Commissures of mitral valve
Pectinate muscles of the right and left atrium/auricle
Left ventricle

Aortic semilunar valve
Nodules
Interventricular septum

IV. Posterior and Superior Mediastinum

Esophagus
Esophageal plexus
Left vagus nerve
Recurrent laryngeal nerve
Ligamentum arteriosum
Left common carotid artery
Subclavian artery
Azygos vein
Intercostal veins
Right vagus nerve
Thoracic lymphatic duct
Descending thoracic aorta
  • Intercostal branches
Sympathetic trunk
Sympathetic chain ganglia
Rami communicantes
Greater splanchnic nerves
Superior mediastinum
Thymus gland
Right and left brachiocephalic veins
Superior vena cava
Brachiocephalic trunk
Left common carotid artery
Left subclavian artery
Phrenic nerves
Tracheobronchial lymph nodes
Carina
Azygos and hemiazygos system of veins

V. Anterior Triangle and Root of the Neck

Deep cervical fascia
Investing fascia
Pretracheal fascia
Prevertebral fascia
Platysma muscle
Sternocleidomastoid muscle
Accessory nerve
Hyoid bone

Thyroid cartilage
Cricoid cartilage
Trachea
Facial vein
Retromandibular vein
Anterior jugular vein
Superior belly of the omohyoid
Inferior belly of the omohyoid
Sternohyoid muscle
Sternothyroid muscle
Thyrohyoid muscle
Cricothyroid membrane
Carotid triangle
- Superior belly of the omohyoid
- Posterior belly of the digastric
- Anterior border of the sternocleidomastoid
Hypoglossal nerve
Carotid sheath
Superior root of the ansa cervicalis
Inferior root of the ansa cervicalis
Vagus nerve
Common carotid artery
Internal carotid artery
External carotid artery
Internal jugular vein
Superior laryngeal nerve
Thyrohyoid membrane
Superior laryngeal artery
Cricothyroid muscle
Internal jugular vein
Superior thyroid artery
Lingual artery
Facial artery
Occipital artery
Carotid sinus region
Carotid body
Superior thyroid vein
Submandibular salivary gland
Mylohyoid muscle
Anterior and posterior bellies of the digastric
Stylohyoid muscle
Thyroid gland
Isthmus
Pyramidal lobe
Superior and inferior thyroid arteries

Recurrent laryngeal nerves
Thoracic duct
Transverse cervical artery
Suprascapular artery
Thyrocervical trunk
Subclavian artery
Vertebral artery
Sympathetic trunk

VI. Posterior Triangle of the Neck

Sternocleidomastoid muscle
Trapezius muscle
Accessory nerve
Subclavian vein
Transverse cervical artery
Suprascapular artery
Scalenus anterior muscle
Thyrocervical trunk
Splenius capitis
Levator scapulae
Scalenus posterior
Scalenus medius
Subclavian artery
Brachial plexus
Phrenic nerve

Block II

REFLECTIONS

Chapter 3

I. Thoracic Wall

How do you view being an anatomical donor?
Would you want to donate your body? Why or why not?
Do you think that anatomical donors have true informed consent?

II. Pleural Cavities and the Lungs

What did your donor's lungs feel like in your hands?
Have you noticed any differences between your donor's lungs, and other donors' lungs?
Think about the translation of the word "phrenic." What do you think it was like to dissect in the nineteenth century?

III. Mediastinum and Heart

What did you anticipate in holding a donor's heart? What did it feel like?
What will you picture when you listen to heart sounds?

IV. Posterior and Superior Mediastinum

What do you think it was like to be a physician in the sixteenth and seventeenth centuries? What did physicians value, and what qualities did physicians need to have?
What are your values as a future physician? What qualities do you think a present-day physician should have?

V. Anterior Triangle and Root of the Neck

Reflect on dissecting the neck.

VI. Posterior Triangle of the Neck

Compare your donor's muscles to the muscles of other donors. What is the same, and what is different?

Do you have a favorite donor to study with aside from your own? Why or why not?

From your perspective, what makes your donor easier or harder to work with?

Post-Exam Reflection

What was the process of taking the exam like for you?

What was it like to stand beside other donors during the exam?

Block III

Chapter 4

HEAD AND NECK

I. Face

Learning Objectives

- If you haven't already, decide as a group when to look at your donor's face.
- Use your atlas to learn the bony landmarks of the face.

A. Introduction

Lifelines as Rooflines

That afternoon in the suites
I know exactly where
You will feel most alive to me

Innate humanity
Lines the palms
Of your curled hands
They do not bring the unease I expected
They cup around my double-gloved ones,
Assenting

A tactile reminder that
Not all homes have four walls
Some are laden with connective tissue
And bound up in a delicately fierce epithelium
Some homes are creased with the imprints of lives touched,
Stiff from braiding strands of sunlight into seven decades of soft breaths

Thank you
For providing these lessons in shelter
A few hours at a time

—Sabrina Sayegh, Class of 2024

B. Bony Landmarks

Maxilla: Latin = jaw.

Zygomatic: Greek zygon = yoke or crossbar that hitches two animals together to draw a plow; a yoke is similar in shape to the bony zygomatic arches on both sides of the skull that form the cheek bones.

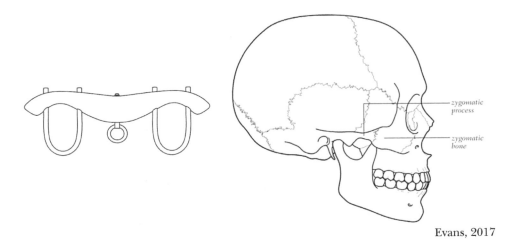

Evans, 2017

Mandible: Latin mandere = to chew.

Occipital: Latin prefix ob- = prominent, and caput = head; hence the prominent convexity of the back of the head.

Frontal: Latin frontis = of the forehead.

Parietal: Latin paries = house wall; any enclosing wall-like structure.

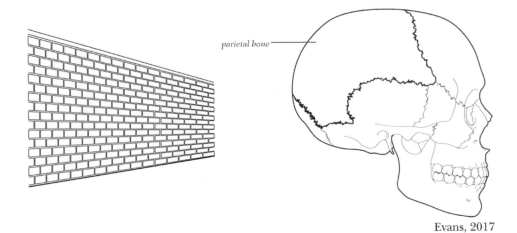

Evans, 2017

Mastoid: Greek mastoeides = breast-shaped.

Coronal: from Latin corona = wreath or crown.

Sagittal: from Latin sagittal = arrow.

Lambdoid: uppercase lambda is the 11th letter of the Greek alphabet.

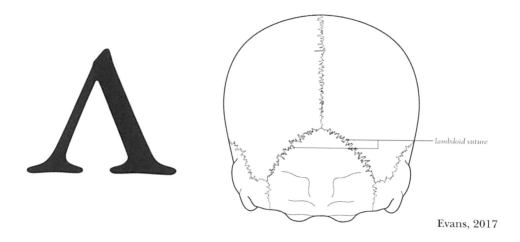

lambdoid suture

Evans, 2017

Bregma: from a Greek word implying moist, referring to the site of the anterior fontanelle, the site of junction of the coronal and sagittal sutures.

Suture: Latin sutura = a seam; the line where two skull bones meet, as in a seam.

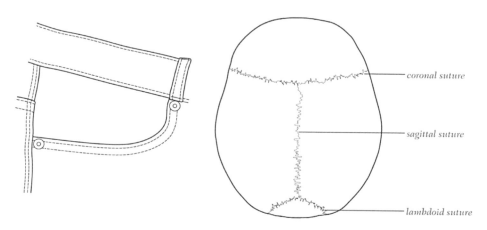

coronal suture

sagittal suture

lambdoid suture

Evans, 2017

Fontanelle: from Latin fons = fountain or little water-spring.

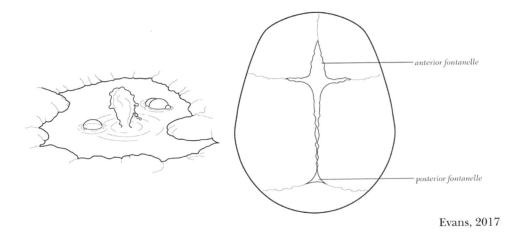

anterior fontanelle

posterior fontanelle

Evans, 2017

II. Interior of Skull and Brain Removal

Learning Objectives

- Think about what it will feel like to hold a human brain in your hands.
- Draw the blood supply to the meninges.
- Draw the dural sinuses, the flow of CSF, and the blood supply to the brain.
- Draw the 12 pairs of cranial nerves on the base of the skull.

A. Introduction

I remember how brutal it felt to saw the pelvis in half. I sawed for 10 minutes straight, not thinking about anything except the task at hand. When it was done I froze, physically and mentally, for over half an hour. I think most people have at least one memory from anatomy lab that is permanently etched into their brains; the Block V split pelvis lab is the one for me.

—Zain Talukdar, Class of 2023

B. Removal of the Calvaria

Aponeurosis: Greek apo = from, and neuron = tendon (later applied to nerve cell and its fibers), used for sheet-like tendons.

Check Your Understanding: draw the layers of the scalp and the sutures of the skull.

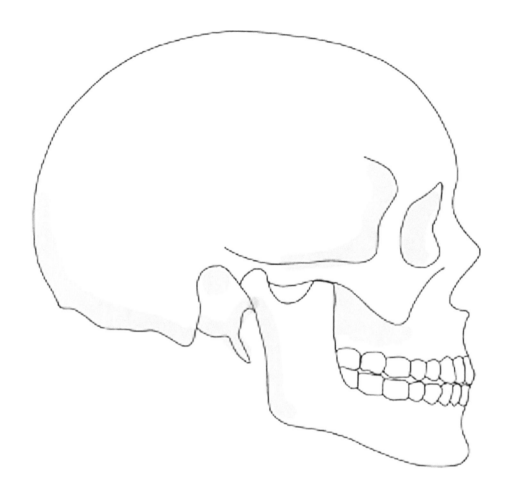

C. Brain Meninges

Meninges: plural of Greek meninx = a membrane.

Mater: Latin = mother.

Dura mater: Latin = tough mother.

Arachnoid mater: derived from Greek arachne = spider; the suffix -oid = similar to.

Pia mater: Latin = tender mother.

Periosteum: Greek peri = around, and osteon = bone; hence, the membrane around a bone.

Lacunae: from Latin lacus = lake.

Sinus: Latin = a recess, bend.

Cerebellum: diminutive of Latin cerebrum = brain.

Check Your Understanding: draw the blood supply to the meninges.

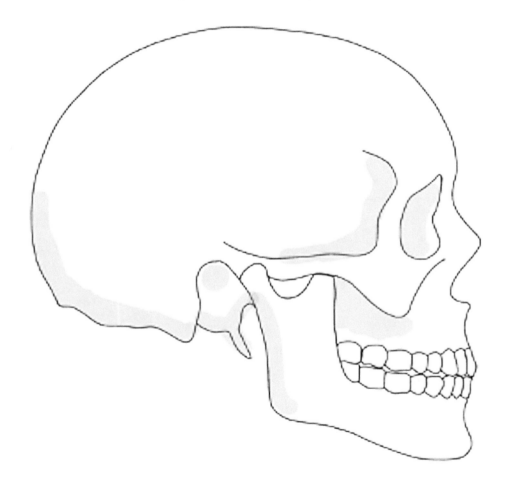

Check Your Understanding: draw the dural sinuses, the ventricles and the flow of CSF.

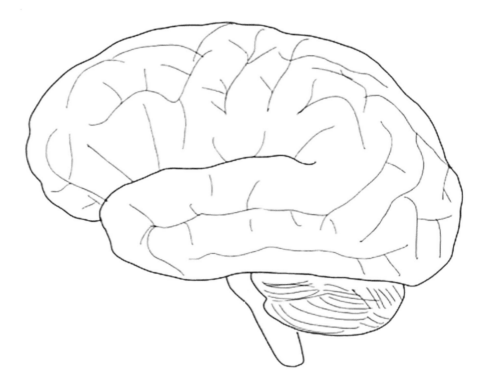

D. Removal of the Brain

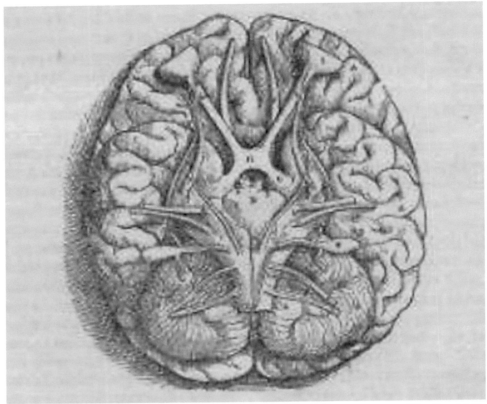

Vesalius, 1543

E. Dural Folds

Tentorium: Latin = tent.

Falx: Latin = a sickle; a curved, serrated, or smooth cutting-tool used for harvesting grain crops.

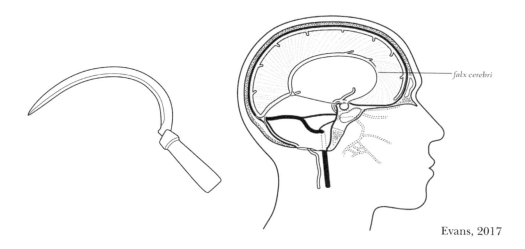

Evans, 2017

Crista galli: Latin crista = crest, and galli = of the cock; hence, a cockscomb.

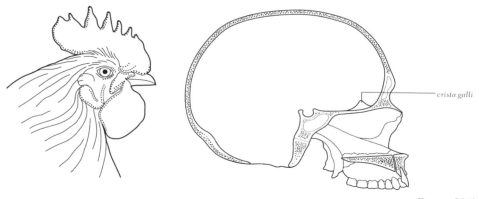

Evans, 2017

Confluens: Latin con = together, and fluens = flowing, hence the meeting of more than one stream.

F. Gross Examination of the Brain

Frontal: from Latin frontis = of the forehead.

Temporal: from Latin tempus = time.

Occipital: from occiput; Latin ob = prominent, and caput = head; hence the prominent convexity of the back of the head.

Parietal: Latin parietalis, pertaining to paries = wall.

Sulcus: Latin = furrow, wrinkle; furrow made in the soil after a field has been plowed.

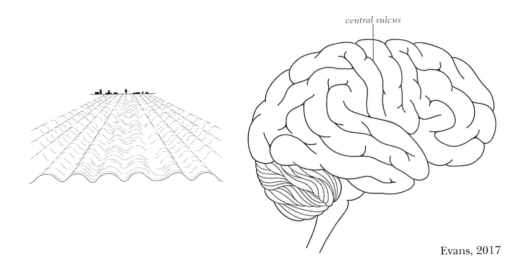

central sulcus

Evans, 2017

Gyrus: Greek guros = a ring; hence, a coil of brain cortex.

G. Cranial Fossae

1. Anterior Fossa

Fossa: from Latin = ditch; hence, an anatomical depression.

2. Middle Fossa

Hypophyseal: Greek adjective; hypo = under, phusis = growth; hence, a undergrowth from the brain.

Pituitary: from Latin pituitarius = secreting phlegm; the gland was thought to produce mucous that discharged through the nose.

Cavernous: Latin = containing caverns or cave-like spaces.

Abducent: Latin abducere = leading away.

Trochlear: from Greek trochilia = a pulley.

Trigeminal: from Latin tri = three, and geminus = twin; hence, the triplets.

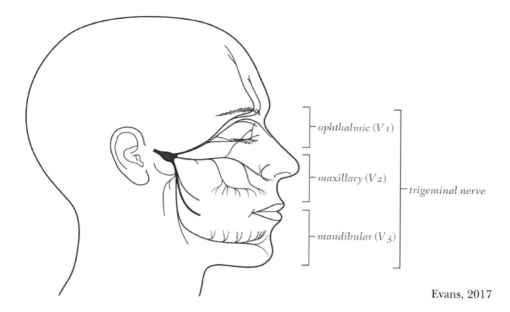

ophthalmic (V_1)

maxillary (V_2)

mandibular (V_3)

trigeminal nerve

Evans, 2017

Ganglion: Greek = tumor on or near sinews or tendons; used by Galen to denote complex nerve centers.

3. Posterior Fossa

Pons: Latin = bridge.

Medulla: Latin = marrow.

Jugular: from Latin jugulum = collarbone, throat.

4. Foramina

Fissure: from Latin findere = to split.

Cribriform plate: from Latin cribrum = sieve, and –iform = in the shape of.

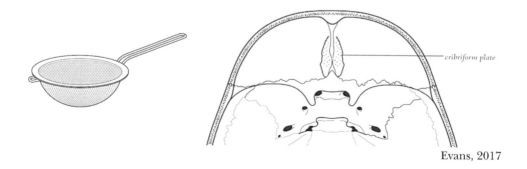

cribriform plate

Evans, 2017

Foramen: from Latin forare = bore a hole.

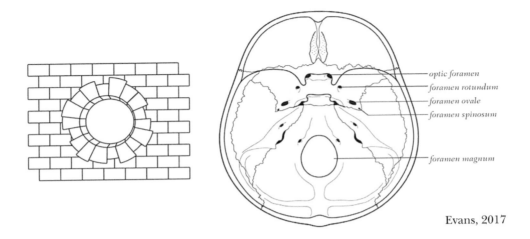

optic foramen
foramen rotundum
foramen ovale
foramen spinosum

foramen magnum

Evans, 2017

Rotundum: from Latin rotundus = round.

Ovale: Latin = oval opening.

Spinosum: named because of its relationship to the spinous process of the greater wing of the sphenoid bone.

Lacerum: Latin = lacerated piercing.

Petrosal: from Latin petrosus = stony, rocky.

Hypoglossal: Latin hypo = under, and glossa = tongue.

Magnum: Latin = great.

Medulla oblongata: Latin medulla = marrow, oblongata = oblong; literally, elongated medulla.

Check Your Understanding: draw the blood supply to the brain and the cranial nerves at the base of the skull.

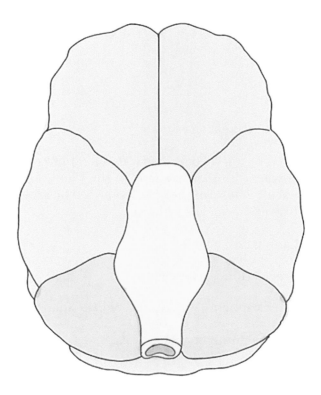

III. Orbit and Eye

Learning Objectives

- Take the time to recognize the reactions of those around you.
- Draw the extraocular muscles.
- Think about what it will be like to look into your patients' eyes with an ophthalmoscope.

(Technical) Manual

Step One: reflect skin flaps back to display
tendons that once moved fingers that held—what?
Did you use a pen, all you had to convey
in clear lines—no mistakes? Did your jaw jut
out as you strained to put words to Step Two:
assembly required? A user guide couldn't
show just how to hold your hand to cut through
each incision marked in Step Three. I wouldn't

have guessed that I'd imagine you scrawling
how-tos with such purpose, setting down text
like you were laying down train tracks—crawling
in an orderly fashion toward the next
stop, where you'd begin Step Four: dissecting
and revising sentences—reflecting.

—Anonymous, Class of 2019

A. Bony Landmarks

Maxillary: from Latin maxilla = jaw-bone.

Zygomatic: Greek zygon = yoke or crossbar that hitches two animals together
to draw a plow; a yoke is similar in shape to the bony zygomatic arches on
both sides of the skull that form the cheek bones. The same word is used in
azygos, where the prefix a- means without.

Lacrimal: from Latin lacrima = tear.

Ethmoid: from Greek ethmos = a sieve.

Sphenoid: from Greek sphen = wedge, and eidos = shape or form.

Palatine: from Latin palatum = palate.

Orbit: from Latin orbis = ring.

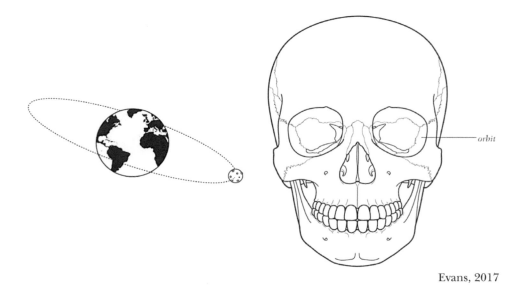

Evans, 2017

B. Right Orbit, Superior Anatomical Approach

Clinoid: from Greek kline = bed, eidos = shape or form; hence, like a bed-post.

Levator palpebrae superioris: from Latin levare = raise, lift; Latin palpebra = eyelid; Latin superus = above.

Trochlear: from Greek trochilia = a pulley.

Oblique: Latin obliquus = slanting or sloping.

Ciliary: from Latin cilium = eyelash.

Rectus: Latin = straight.

Check Your Understanding: draw the extraocular muscles.

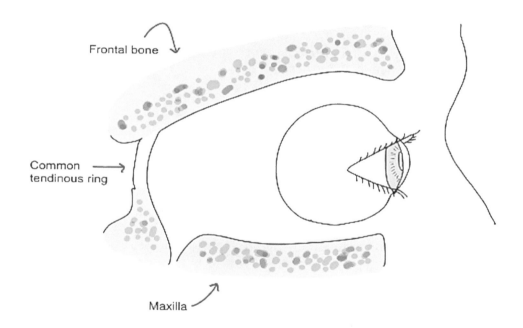

C & D. Left Orbit: Surgical Approach and Dissection of Globe

Sclera: from Greek skleros = hard.

Cornea: from Latin cornu = horn.

Choroid: from Greek chorion = skin, and eidos = shape or form; hence, like a membrane.

Papilla: from Latin papula = small protuberance.

Lens: from Latin lentil; named because of the similarity in shape.

Pupil: from Latin pupa = doll; named for the tiny reflected images visible in the eye.

IV. Temporal Region

Learning Objectives
- Think about the time you have spent in lab since your anatomy course started.

A. Bony Landmarks

Temporal: from Latin tempus = time.

Styloid: Greek stylos = an instrument for writing, and eidos = shape or form; hence a pen- or pencil-like structure.

Meatus: Latin = passage.

Coronoid: from Greek korone = a crown, eidos = shape or form; hence, shaped like a crown.

Lingula: diminutive of Latin lingua; hence, a little tongue.

Pterygopalatine: relating to the pterygoid process and the palatine bone.

Sphenoid: from Greek sphen = wedge, and eidos = shape or form.

B & C. Preparation of the Dissection Field and Masseter Muscle

Masseter: from Greek masasthai = to chew.

Temporalis: from Latin tempus = time.

D. Infratemporal Fossa

Alveolar: from Latin alveus = a small cavity.

Lingual: Latin lingua = tongue.

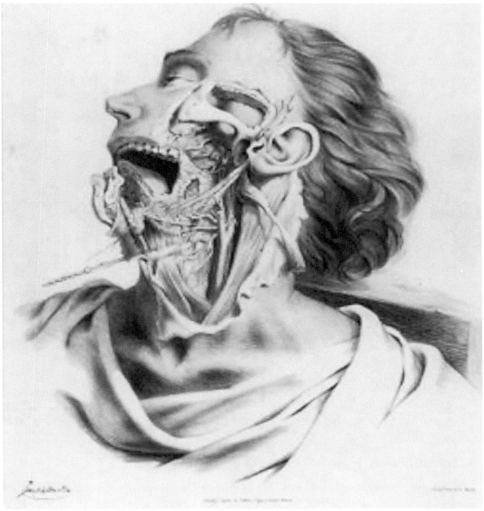

Quain, 1844

Chorda tympani: Latin chorda = string, tympanum = drum; so named because it crosses the ear drum in the middle ear.

Auriculotemporal: of or relating to the auricle of the ear and the temple.

Pterygoid: Greek pteryx = wing, and eidos = shape or form; hence, wing or feather-shaped.

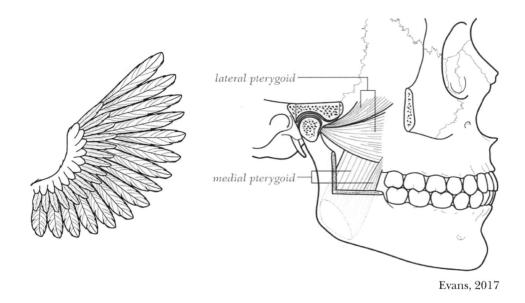

Evans, 2017

V. Retropharyngeal Region

Learning Objectives

- Reflect on disarticulating the head.
- Think about the ethics of head transplants.

A. Introduction

"So that's the spring ligament? We're done? Like done done? This early??"
Closing my donor, as I had done at least fifty times before, it hit me. I froze. I'd sprayed the Downey solution, I was holding a wet paper towel. I couldn't bring myself to close the skin and wrap the twine, not just yet. I was thinking about my first-day pep talk (you're fine, this is your new normal, please do not pass out, here we go) and the roller coaster ride that followed.

—Sara Peterson, Class of 2022

B. Bony Landmarks

Atlas (C1): named for the Atlas of Greek mythology, who was condemned to hold up the sky for eternity. Likewise, the atlas supports the globe of the head.

Axis (C2): axle or pivot; the pivot around which the first cervical vertebra, the atlas, rotates.

Dens: Latin = tooth.

Odontoid: Greek odous = tooth, and eidos = form, shape; hence, tooth-like.

Articular: from Latin articulus = small connecting part.

Condyle: from Greek kondylos = knuckle.

C. Craniovertebral Joints

Tectorial: from Latin tectorium = covering, a cover.

Cruciform: from Latin crux = cross.

Alar: from Latin ala = wing.

D & E. Disarticulation of the Head and Neck and Pre- and Lateral Vertebral Regions

Longus colli: Latin longus = long, and Latin collum = neck.

Scalenus: from Latin scalenus musculus = unequal muscle.

F. Base of the Skull.

Nodose: from Latin nodus = knot.

Laryngeal: relating to the larynx; Greek larynx = voice-box.

VI. Pharynx

Learning Objectives

- Think about the pros and cons of dissecting donors instead of looking at prosected donors or using visual software.

A. Introduction

I don't believe this; I think it is self-involved to say that we are doing this for the greater good.

If it was just the greater good, we would cremate your brain with your body.

The professor tells me that only the bone remnants survive the fire and occupy the ash that you will become. Your brain isn't bone. So therefore, it doesn't matter, anyway?

But yet, you will walk into your next life with no left eyelid, a little less adipose tissue, and no brain. Not even a right prefrontal cortex.

A dog never ceases to be a dog once it dies. It just becomes a dead dog. You were Stella. You still are a person. Your personhood doesn't cease with the activity of your cells.

—Anonymous, Class of 2022

B. External Pharynx

Constrictor: Latin con = together, and strictum = bound tightly; hence, producing narrowing.

Thyroid: from Greek thureoeides = shield shaped.

Cricoid: from Greek krikos = a ring, the suffix -oid = similar to; ring-like.

Recurrent: from Latin recurrere = running back; hence, a structure that bends and runs back toward its source.

C. Internal Pharynx

Nasal choanae: Greek choana = funnel.

Septum: Latin septum = a fence, or boundary wall.

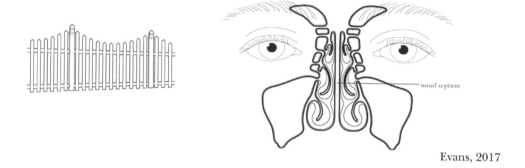

Evans, 2017

Torus tubarius: Latin torus = bulge.

Palatine: from Latin palatum = palate.

Tonsil: Latin tonsilla = tonsil.

Epiglottis: Greek epi = upon, near to, and glotta = tongue.

Piriform recess: from Latin pirum = a pear; hence, pear-shaped; recess = a secluded area or pocket; hence, a small cavity set apart from a main cavity.

VII. Tongue and Nasal Cavity

Learning Objectives

- Observe the mouths of different donors.
- Think about your own eating habits.

A. Introduction

Over the past 14 weeks, while many of us derived comfort from imagining the lives our donors might have led, we ultimately understand that we will never know these 26 people. We will never get to know those who gave us this incredible privilege as we took our first steps into the medical profession. No doubt, our donors collectively represent many walks of life. Beyond their names, ages, and occupations, we know little of their lives. We will never know how they spent their Saturday mornings, what hobbies filled their free time, which causes they felt passionately about, where they grew up, which loved ones' photos they carried in their wallets, whether they found faith in religion, how they spent their final days and moments.

—Camille Corre, Class of 2023

B. Tongue

Sulcus terminalis: from Latin sulcus = furrow, wrinkle; furrow made in the soil after a field has been plowed; terminalis = terminal, marking a boundary.

Foramen cecum: from Latin forare = bore a hole; from Latin caecum = blind.

Fungiform papillae: fungiform = having the shape of or resembling a fungus or mushroom; from Latin papula = small protuberance.

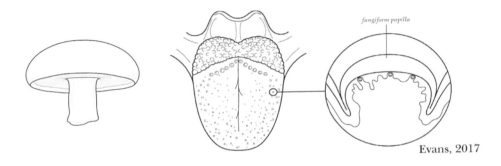

Evans, 2017

Filiform papillae: filiform = threadlike; Latin papula = small protuberance.

Circumvallate papillae: Latin circum = around, and vallum = wall; Latin papula = small protuberance.

Valleculae: from Latin valles = valley.

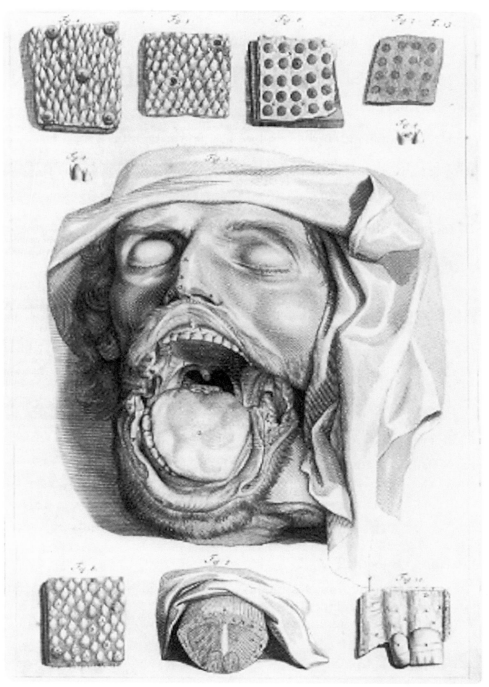

Bidloo, 1690

C. Nasal Cavities

Incisive: from Latin incindere = cut into.

Lacerum: from Latin = lacerated piercing.

Concha: from Latin conch = shell.

Turbinate: from Latin turbin = spinning top, whirl.

Vomer: Latin = plowshare: the main cutting blade of a plow.

D. Lateral Wall of the Nasal Cavity

Meatus: Latin = passage.

Hiatus: from Latin haire = gape.

Ostium: Latin = door.

Infundibulum: Latin = funnel.

Infra-orbital: Latin infra = below, and Latin orbis = ring.

E. Pterygopalatine Ganglion

Palatine: from Latin palatum = palate.

VIII. Palate, Mouth, and Nasopharyngeal Wall

Learning Objectives
- Many of the muscles in this lab are named by their inferior and superior attachments.
- Understanding the names of muscles can help you identify them.

A. Palatine Region

The following structures are named by their inferior and superior attachments:

Palatopharyngeus: relating to the palate and the pharynx.

Stylopharyngeus: relating to the styloid process and the pharynx.

Glossopharyngeal: relating to the tongue and the pharynx.

Styloglossus: relating to the styloid process and the tongue.

B. Sublingual Region

Maxilla: Latin = jaw.

Mandible: from Latin mandere = to chew.

Masseter: from Greek masasthai = to chew.

Frenulum: from Latin frenum = curb.

Parotid: from Greek para = bedside, and otos = of the ear; hence, beside the ear.

Plica: from Latin plicare = to fold; hence, a fold.

Sublingual: Latin sub = under or below, and lingua = tongue; hence, under the tongue.

Hamulus: from Latin hamus = hook.

Submandibular: beneath the jaw or mandible.

The following structures are named by their inferior and superior attachments:

Mylohyoid: Greek mylo = molar, and hyoeides = U-shaped.

Geniohyoid: Greek genion = chin, and hyoeides = U-shaped.

Genioglossus: Greek genion = chin, and glossa = tongue.

Hypoglossal: Greek hypo = under, and glossa = tongue.

Hyoglossus: Greek hyoeides = U-shaped, and glossa = tongue.

C. Tongue

The following structures are named by their inferior and superior attachments:

Styloglossus: relating to the styloid process and the tongue.

Palatoglossus: relating to the palate and the tongue.

Genioglossus: relating to the chin and the tongue.

D. Nasopharyngeal Wall

Palatopharyngeus: relating to the palate and the pharynx.

Buccinator: Latin = trumpeter; hence, the muscle that blows air out from the cheek under pressure.

Levator veli palatini: Latin levare = to lift, vellum = veil or curtain, and platum = the roof of the mouth; hence, the muscle that lifts the veil, or curtain, at the roof of the mouth.

Tensor veli palatini: Latin tendere = to stretch, vellum = veil or curtain, and platum = the roof of the mouth; hence, the muscle that stretches the veil, or curtain, at the roof of the mouth.

IX: Larynx

Learning Objectives

* Think about what your donor's voice was like.

A. Laryngeal Cartilages

Cricoid: ring-like; Greek krikos = a ring, the suffix -oid = similar to.

Thyroid: from Greek thureoeides = shield-shaped.

Arytenoid: from Greek arytaina = pitcher, and eidos = shape or form; arytenoid cartilage curves like a spout.

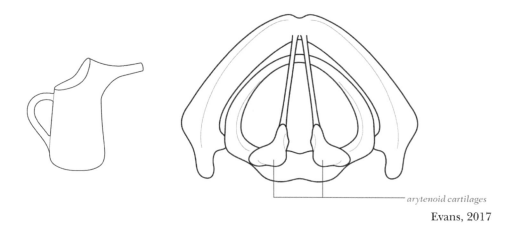

arytenoid cartilages

Evans, 2017

B & C. Laryngeal Muscles and Interior of the Larynx

Vestibule: a partly enclosed space in front of the entrance to a Roman house.

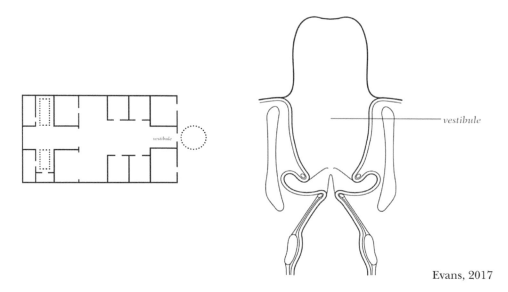

vestibule

Evans, 2017

Ventricle: from Latin ventriculus, diminutive of venter = belly.

Infraglottic: Latin infra = below; hence, below the glottis. Vocal: Latin vox = voice.

X. Ear

A. Introduction

For Kathleen

The truth is,
Two roads diverged and
I wasn't sure if I chose the right one
Then I met you at my crossroads
The truth is,
Were you person, body, or object?
You were a salesperson and now you are
A scaffold for the education of future healers

The truth is,
You are the first drop of a hurricane.
Beauty, strength, and intelligent design -
A legacy that builds beyond you.

The truth is,
there is still so much I don't know
But with each cut, I slice away at
Doubts amassed along my path to you

The truth is,
I found purpose within the folds of your skin.
Each cut a quest to understand how you came to be,
A journey to illuminate mysteries of the human body.

—Michelle Duan, Class of 2024

B. Outer Ear

Meatus: Latin = a passage or channel, or its external structure.

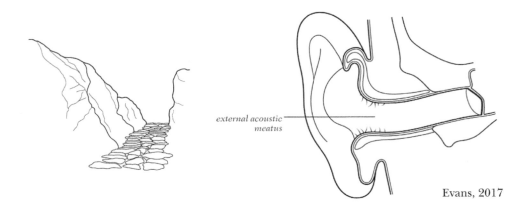

external acoustic
meatus

Evans, 2017

C. Middle Ear

Malleus: Latin = hammer.

Incus: Latin = anvil; hence, the anvil-shaped ossicle of the middle ear.

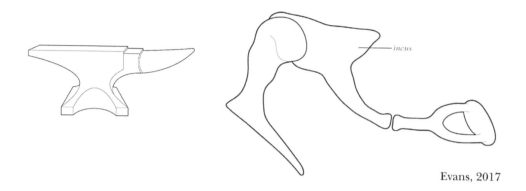

Evans, 2017

Stapes: Latin = stirrup.

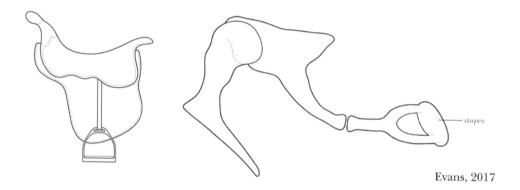

Evans, 2017

Fenestra: Latin = a window; hence, a small hole or opening in a bone.

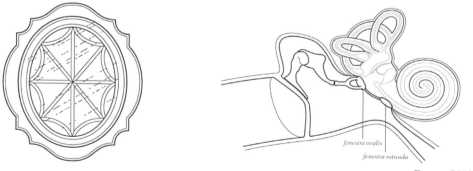

Evans, 2017

Petrous: from Latin petrosus = rocky, stony.

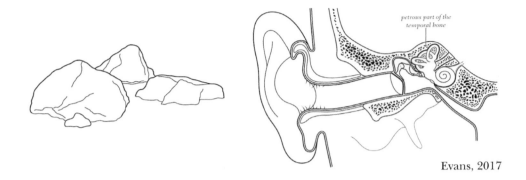

Evans, 2017

D. Inner Ear

Tympanum: Latin = a drum.

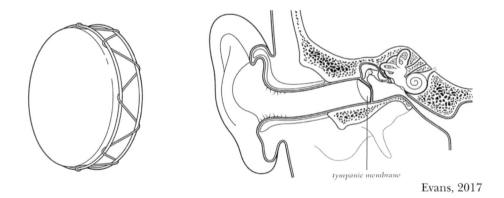

Evans, 2017

Cochlea: from Greek kokhlias = snail; hence, a spiral shell.

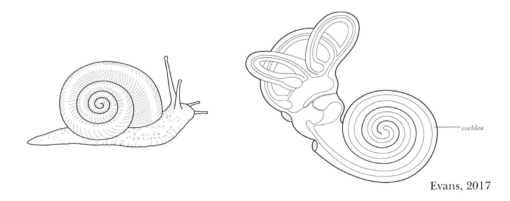

Evans, 2017

Scala: Latin = a staircase.

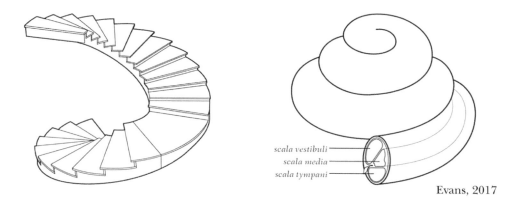

scala vestibuli
scala media
scala tympani

Evans, 2017

Utricle: from Latin utriculus = a small leather bag or bottle.

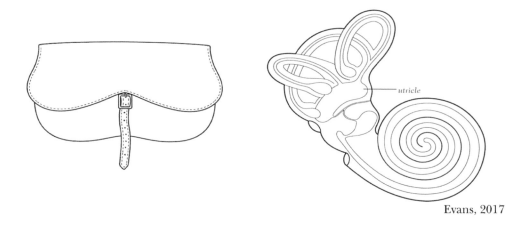

utricle

Evans, 2017

Saccule: from Latin sacculus = a small sac or pouch.

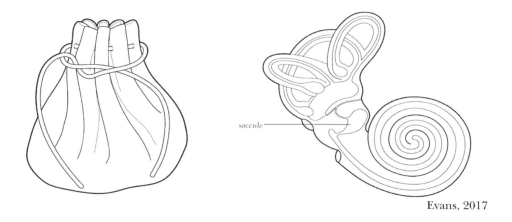

saccule

Evans, 2017

Tectorial membrane: from Latin tectum = a roof.

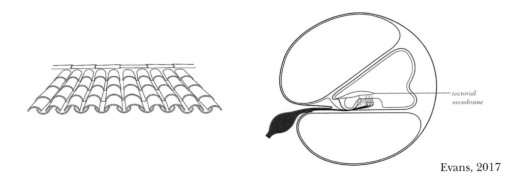

tectorial
membrane

Evans, 2017

Block III

BOLD TERMS

I. Face

Frontal bone
Maxilla
Zygomatic bone
Mandible
Teeth
Lacrimal bone
Coronal suture
Parietal bones
Sagittal suture
Bregma
Occipital bone
Lambdoid sutures
External auditory meatus
Temporal bone
Zygomatic arch
Mastoid process
Styloid process
Stylomastoid foramen

II. Interior of Skull and Brain Removal

Layers of the scalp
Brain meninges
Dura mater
Periosteal (endosteal) dura
Meningeal layer
Middle meningeal artery
Superior sagittal dural sinus
Right and left transverse dural sinuses
Lacunae laterales
Arachnoid granulations
Superior sagittal sinus
Cerebellar hemispheres

Arachnoid mater
Pia mater
Dural folds
Cerebellar tentorium
Cerebellar falx
Cerebral falx
Cerebral hemispheres
Crista galli
Superior sagittal sinus
Inferior sagittal dural sinus
Straight sinus
Confluence of the sinuses
Frontal lobes
Temporal lobes
Occipital lobes
Parietal lobes
Lateral sulcus
Central sulcus
Vertebral arteries
Posterior inferior cerebellar arteries
Basilar artery
Anterior inferior cerebellar arteries
Superior cerebellar arteries
Oculomotor nerve
Posterior cerebral arteries
Posterior communicating arteries
Internal carotid arteries
Cerebral arterial circle
Middle cerebral artery
Anterior cerebral arteries
Anterior communicating artery
Twelve pairs of cranial nerves
Trochlear nerve
Cranial fossae
Anterior fossa
Sphenoid
Ethmoid
Frontal bones
Middle fossa
Temporal bones
Hypophyseal fossa
Pituitary gland
Cavernous sinus
Abducent nerve
Oculomotor nerve

Trochlear nerve
Trigeminal nerve
Internal carotid artery
Trigeminal ganglion
Divisions of cranial nerve V
- V1: ophthalmic division
- V2: maxillary division
- V3: mandibular division

Posterior fossa
Pons
Medulla
Internal acoustic meatus
Jugular foramen
Hypoglossal canal
Foramina
- Anterior cranial fossa
  - Cribriform plate
  - Olfactory nerves
- Middle cranial fossa
  - Optic canal
    - Optic nerve
    - Ophthalmic artery
    - Superior orbital fissure
    - Oculomotor nerve
    - Trochlear nerve
    - V1
    - Abducent nerve
    - Ophthalmic veins
  - Foramen rotundum
    - V2
  - Foramen ovale
    - V3
    - Lesser petrosal nerve
  - Foramen spinosum
    - Middle meningeal artery
  - Foramen lacerum
  - Carotid canal
    - Internal carotid artery and nerve plexus
- Posterior cranial fossa
  - Internal acoustic meatus
    - Facial nerve
    - Vestibulocochlear nerve
- Jugular foramen
  - Glossopharyngeal nerve
  - Vagus nerve

- Accessory nerve
- Inferior petrosal sinus
- Sigmoid sinus
- Internal jugular vein
- Hypoglossal canal
  - Hypoglossal nerve
- Foramen magnum
  - Medulla oblongata
  - Spinal roots of accessory nerve
  - Vertebral arteries

III. Orbit and Eye

Maxillary bone
Zygomatic bone
Frontal bone
Lacrimal bone
Ethmoid bone
Sphenoid bone
Palatine bone
Optic canal
Superior orbital fissure
Greater wing of sphenoid bone
Lesser wing of sphenoid bone
Inferior orbital fissure
Infra-orbital groove
Infra-orbital canal
Infra-orbital foramen
Anterior ethmoidal foramina
Posterior ethmoidal foramina
Cribriform plate
Periorbita
Frontal air sinus
Anterior and posterior ethmoid air cells
Anterior clinoid process
Levator palpebrae superioris muscle
Trochlear nerve
Superior oblique muscle
Frontal nerve
Supratrochlear nerve
Supraorbital nerve
Lacrimal nerve
Lacrimal gland
Superior rectus muscle
Oculomotor nerve

Superior oblique muscle
Lateral rectus muscle
Abducent nerve
Nasociliary nerve
Long ciliary branches
Oculomotor nerve
Ciliary ganglion
Short ciliary nerves
Inferior rectus muscle
Inferior oblique muscle
Superior ophthalmic vein
Cavernous sinus
Ophthalmic artery
Internal carotid artery
Common tendinous ring
Sclera
Cornea
Choroid
Ciliary body
Iris
Retinal layer
Optic papilla
Lens
Pupil
Optic nerve
Lacrimal fossa

IV. Temporal Region

Temporal bone
Styloid process
External acoustic meatus
Mandibular fossa
Mandible
- Head
- Neck
- Ramus
- Angle
Mandibular notch
Lingula
Mandibular foramen
Mylohyoid line
Temporal fossa
Zygomatic arch
Lateral pterygoid plate of the sphenoid

Infratemporal surface of the maxilla
Pterygopalatine fossa
Lateral pterygoid plate
Greater wing of the sphenoid
Foramen ovale
Foramen spinosum
Parotid duct
Facial nerve
Masseter muscle
Temporalis muscle
Temporomandibular joint (TMJ)
Inferior alveolar nerve and artery
Mylohyoid nerve
Mental foramen
Mental nerve
Lingual nerve
Lateral pterygoid muscle
Maxillary artery
Lateral pterygoid muscle
Chorda tympani nerve
Medial pterygoid muscle
Auriculotemporal nerve
Middle meningeal artery
Articular disc

V. Retropharyngeal Region

Retropharyngeal (retrovisceral) space
Axis
Dens/odontoid process
Atlas
  • Posterior arch
  • Anterior arch
  • Transverse process
  • Superior articular facet
Occipital bone
Foramen magnum
Occipital condyles
Tectorial membrane
Cruciform ligament
Alar ligaments
Superior cervical ganglia
Longus colli muscle
Longus capitis muscle
Scalenus anterior muscle

Vertebral artery
CN IX
CN X
CN XI
CN XII
Inferior ganglion (nodose ganglion)
Superior laryngeal nerve

VI. Pharynx

Buccopharyngeal fascia
Middle constrictor
Superior constrictor
Inferior constructor
Thyroid cartilage
Cricoid cartilage
Internal laryngeal branch of superior laryngeal nerve
Thyrohyoid membrane
Recurrent laryngeal nerve
Stylopharyngeus muscle
Nasopharynx
Nasal choanae
Nasal septum
Torus tubarius
Oropharynx
Palatoglossal arch
Palatophargyngeal arch
Palatoglossus
Palatopharyngeus
Palatine tonsil
Laryngopharynx (hypopharynx)
Epiglottis
Piriform recess
Internal laryngeal nerve
Recurrent laryngeal nerve

VII. Tongue and Nasal Cavity

Sulcus terminalis
Foramen cecum
Fungiform papillae
Filiform papillae
Circumvallate papillae
Median glossoepiglottic fold
Valleculae

Lingual tonsils
Geniohyoid
Palatine process of maxilla
Horizontal plates of palatine bones
Incisive foramen
Greater and lesser palatine foramina
Perpendicular plate of the palatine bone
Sphenopalatine (pterygopalatine) foramen
Pterygopalatine fossa
Sphenoid sinus
Pterygoid canal
Foramen lacerum
Sphenopalatine foramen
Frontal process of the maxilla
Inferior concha (turbinate)
Middle and superior conchae
Maxillary sinus
Nasolacrimal canal
Frontal sinus
Cribriform plate
Nasal Septum

- Vomer
- Perpendicular plate of the ethmoid bone
- Septal cartilage

Nasopalatine nerve
Incisive canal
Superior conchae
Middle conchae
Inferior conchae
Auditory tube
Nasolacrimal duct
Inferior meatus
Hiatus semilunaris
Ostium for the maxillary sinus
Ethmoidal bulla
Frontal sinus
Infundibulum
Posterior ethmoid air cells
Sphenoethmoidal recess
Infra-orbital canal
Infra-orbital nerve and vessels
Greater palatine nerve
Sphenopalatine foramen
Pterygopalatine ganglion
Sphenopalatine artery

Greater palatine nerve and vessels
Nerve of the pterygoid canal

VII. Palate, Mouth, and Nasopharyngeal Wall

Palatopharyngeus muscle
Superior constrictor muscles
Stylopharyngeus muscle
Glossopharyngeal nerve
Styloglossus muscle
Maxilla
Zygomatic arch
Ramus of the mandible
Coronoid process
Masseter muscle
Frenulum
Orifice of the parotid duct
Lingual frenulum
Opening of the submandibular duct
Plica sublingualis
Hamulus
Medial pterygoid plate
Mylohyoid line
Sublingual fossa
Mylohyoid muscle
Geniohyoid muscle
Genioglossus muscle
Sublingual gland
Submandibular salivary gland
Submandibular duct
Lingual nerve
Submandibular ganglion
Hypoglossal nerve
Hyoglossus muscle
Lingual artery
Extrinsic tongue muscles
- Hyoglossus
- Styloglossus
- Palatoglossus
- Genioglossus
Nasopharyngeal wall
Palatoglossus muscle
Palatopharyngeus muscle
Buccinator muscle
Opening of the auditory (eustachian) tube

Levator veli palatini muscle
Tensor veli palatini muscle
Scaphoid fossa
Hamulus
Soft palate
Cricoid cartilage
Thyroid cartilage
Arytenoid cartilage
Posterior cricoarytenoid muscle
Arytenoideus
Cricothyroid muscle
Vestibule
Ventricle
Infraglottic cavity
Vestibular (false) folds
Vocal (true) folds or cords
Vocal ligament (fold)

Block III

REFLECTIONS

Chapter 4

I. Face

What did you feel when you held the skull in your hand?

II. Interior of Skull and Brain Removal

How did it feel to view the brain being removed from the skull?
What did it feel like to hold the brain in your hands? Was it what you expected?
What do you think it will be like to do a neurological exam with your patients?

III. Orbit and Eye

Did you observe your group members during this lab?
How did you support one another through this dissection?
What will you picture when you do an eye exam with your patients?

IV. Temporal Region

How long does it feel like you have been in medical school?
How long does it feel like you've been in anatomy lab?
What makes the time go faster or slower?
How much time do you think physicians should spend with their patients?

V. Retropharyngeal Region

Think about head transplants. Who should decide if head transplants are ethical or not?

VI. Pharynx

What is the point of all of this? Do you think physicians need to dissect a donor?

What does being a doctor mean to you?

VII. Tongue and Nasal Cavity

Have your eating habits changed since you started medical school?
What are some similarities and differences that you have noticed between the mouths of different donors?

VIII. Palate, Mouth, and Nasopharyngeal Wall

Reflect on dissecting the mouth.

IX. Larynx

What do you think your donor's voice was like?
What type of family do you imagine your donor had?
What are your thoughts on patient and family-centered medicine, instead of patient-centered medicine?

X. Ear

Reflect on dissecting the ear.

Post-Exam Reflection

What was the process of taking the exam like for you?
What was it like to stand beside other donors during the exam?

Block IV

CHAPTER 5

Abdomen

I. Anterior Abdominal Wall

A. Landmarks and Surface Anatomy

Xiphisternal junction: articulating with the xiphoid, and the sternum.

Pubic: from Latin os pubis = the bone of the pubes.

Symphysis: Greek syn = together, and physis = growth; hence, growing together, or a joint where union between the bones is by fibrocartilage.

Iliac: from Latin ilia = the bone of the flank.

Inguinal: from Latin inguen = groin.

B. Muscles of the Anterior Wall: External Oblique

Oblique: Latin obliquus = slanting or sloping.

Transversus abdominis: Latin transversus = turn across, and abdomen = the belly, the part of the trunk between the thorax and the perineum.

Dartos: Greek = flayed or skinned.

Epigastric: Greek epi = upon, and gaster = belly.

Linea alba: Latin linea = line, and alba = white.

C. Inguinal Region

Spermatic: Greek sperma = seed.

Crus: Latin = leg.

Canal: from Latin canalis = an artificial waterway.

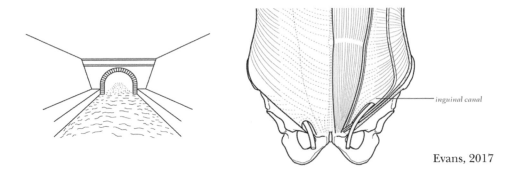

inguinal canal

Evans, 2017

Intercrural: from Latin inter = between, and crus = leg; hence, between leg-like structures.

Lacunar: from Latin lacus = lake.

II. Scrotum, Spermatic Cord, and Testis

Learning Objectives

- Draw the distinguishing features of a direct and an indirect hernia.
- Draw the contributions of the anterior abdominal wall to the layers of the spermatic cord/testis.

A. Introduction

I hold you. I trace your triceps in my fingers. I cup your cheek in my palm. I stroke your nerves. You gifted me the knowledge of your body. Intimate knowledge. A knowing of you that you might not even have had yourself. You've gifted me a piece of your humanness, of your personhood, of your very being that I've swallowed with the particles of formaldehyde, absorbed through the slickness that moisturizes my gloves. You will always live on inside me. You live in me until I die, and until those I touch die. You're a part of my healing touch now. You taught me to be gentle. To cut with precision and to inflict only the necessary change to reveal your flowing, interconnected structures. You've given me your body, and knowledge, and now, I'm realizing, a bit of your wisdom.

–Anonymous, Class of 2022

B. Scrotum and Spermatic Cord

Scrotum: from Latin scorteus = leather, a hide.

Cremaster: from Greek krema = hang or suspend.

Ductus (Vas) deferens: Latin vas = vessel, ductus = a duct, deferens = carrying down.

Pampiniform: Latin = in the shape of a young vine shoot.

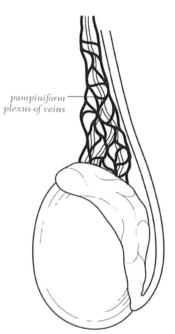

pampiniform plexus of veins

Evans, 2017

C. Testis

Testis: Latin = a witness. Under Roman law, no man could bear witness (testify) unless he possessed both testes.

Tunica albuginea: Latin *tunica* = a shirt or garment; Latin *albus* = white.

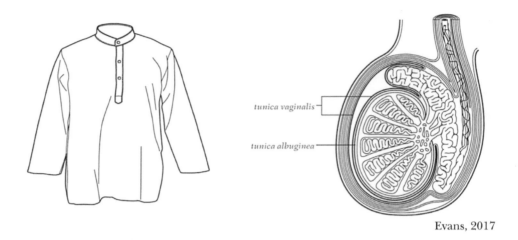

Evans, 2017

Epididymis: Greek *epi* = upon, and *didymos* = testis; hence, the organ perched posterosuperior to the testis.

Rete: Latin = a net, snare, or network.

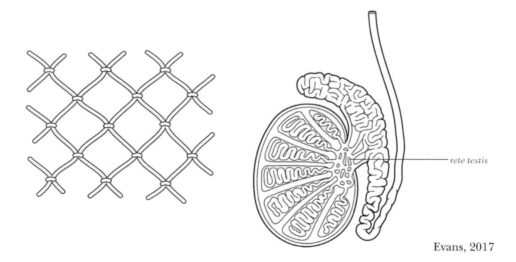

Evans, 2017

D. Muscles of the Anterior Wall Continued

Rectus: Latin = straight.

Arcuate: from Latin arcuatum = curved or arched.

Check Your Understanding: draw the distinguishing features of a direct and indirect hernia.

III. Peritoneum and Peritoneal Cavity

Learning Objectives
- Reflect on dissecting the peritoneal cavity.
- Think about your own eating habits since you started medical school.

A. Orientation

Visceral: Latin viscus = internal organ.

Parietal: Latin parietalis, pertaining to paries = wall.

Peritoneum: Greek periteino = to stretch around; hence, the membrane stretched around the internal surface of the walls and the external aspect of some of the contents of the abdomen.

Mesenteries: Greek mesos = middle, and enteron = intestine; hence, the peritoneal fold that tethers the centrally situated small intestine.

Falciform ligament: from Latin falx = a sickle; a curved, serrated, or smooth cutting-tool used for harvesting grain crops.

Teres: Latin = rounded, cylindrical.

Ligament: Latin ligamentum = bandage.

Umbilical: Latin umbilicus = navel.

B. Gastrointestinal Tract

Stomach: Greek stomachos = gullet or oesophagus; later applied to the wider part of the digestive tract, just below the diaphragm.

Fundus: Latin = bottom or base (note that the fundus of the stomach and the uterus are at the top).

Antrum: Greek antron = a cave; hence, a space in a bone or organ.

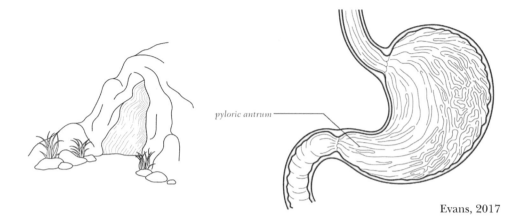

pyloric antrum

Evans, 2017

Pylorus: Greek = gate-keeper; hence, the part of the pyloric canal containing the sphincter, which guards the opening into the duodenum.

Hepatogastric: Greek, hepar = liver, and gaster = belly; hence, relating to the liver and the stomach.

Omentum: Latin = apron.

Duodenum: Latin duodeni = in twelves; the length of the first portion of the small intestine is said to be equivalent to the breadth of approximately twelve fingers.

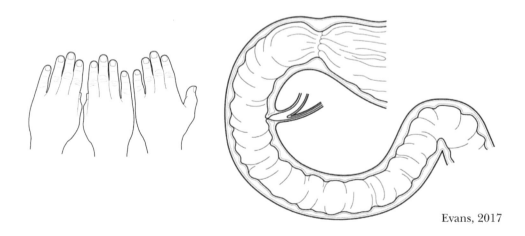

Evans, 2017

Jejunum: Latin jejunus = fasting (because it is found to be empty after death).

Ileum: Greek eilein = twisted.

Cecum: from Latin caecus = blind.

Vermiform appendix: Latin vermis = worm, forma = shape, apprendere = to hang on.

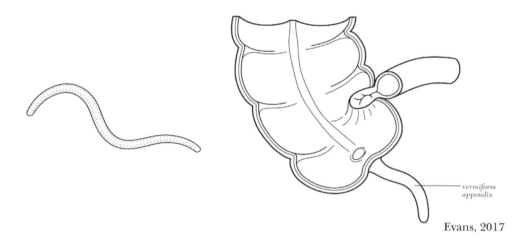

vermiform appendix

Evans, 2017

Colon: Greek kolon = large intestine; hollow.

Flexure: from Latin flexura = a bending.

Phrenicolic: Greek, phren = diaphragm; kolon = large intestine, hollow.

Mesocolon: the mesentery of the colon.

Rectum: Latin = straight; the rectum was named from animals in which the rectum is straight. It is not straight in humans.

Sigmoid: lowercase sigma is the 18th letter of the Greek alphabet.

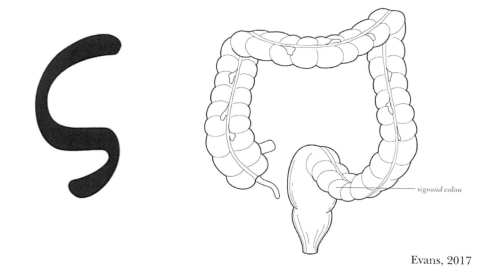

sigmoid colon

Evans, 2017

Taeniae (tenia/teniae): a tape or ribbon.

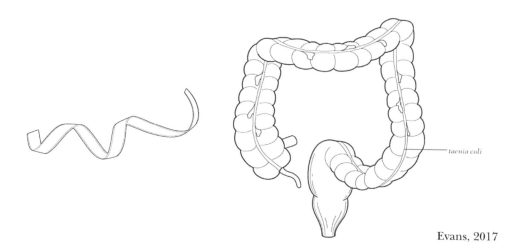

taenia coli

Evans, 2017

Haustra: Latin haustrum (singular) = a scoop or bucket used in drawing water, as in an irrigation system or a water pump; like the sequential filling of sacks or buckets in an ancient irrigation system, one haustrum fills, and distends. This stimulates muscles to contract, pushing the colonic contents onto the next haustrum.

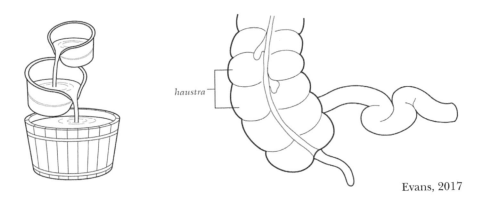

haustra

Evans, 2017

Epiploic: Greek epiploon = a net, which the greater omentum resembles with fat entangled in it.

C. Omental Bursa and Peritoneal Reflections

Bursa: Greek = a purse; hence, a flattened sac containing a film of fluid.

Pancreas: from Greek pancreas, pan = all, and kreas = flesh.

IV. Bile Passages, Celiac Trunk, and Portal Vein

Learning Objectives

- Draw the branches of the celiac trunk.
- Draw the common bile duct and its bifurcation.
- Draw the hepatic vasculature.

A. Introduction

. . . But when we entered anatomy lab for the first time
and made the first incisions peeling back skin,
tentatively removing tiny globs of fat
revealing the trapezius and latissimus dorsi muscles

searching for the spaghetti-like accessory nerve
your body did not feel like a house— perhaps a haunted one.
—Kate Crofton, Class of 2021

B. Common Bile Duct

Hepatoduodenal: relating to the liver and the duodenum.

Common: formed of or dividing into two or more branches.

Cystic: Latin = referring to the gallbladder.

Check Your Understanding: draw the branches of the celiac trunk.

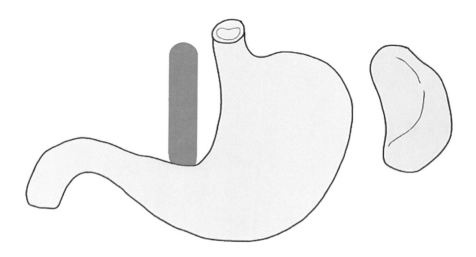

C. Hepatic Artery

Proper: from Latin proprius = one's own, particular to itself; hence, not dividing into branches.

Gastroduodenal: relating to the stomach and the duodenum.

Celiac: from Greek koilia = belly.

Check Your Understanding: draw the hepatic vasculature.

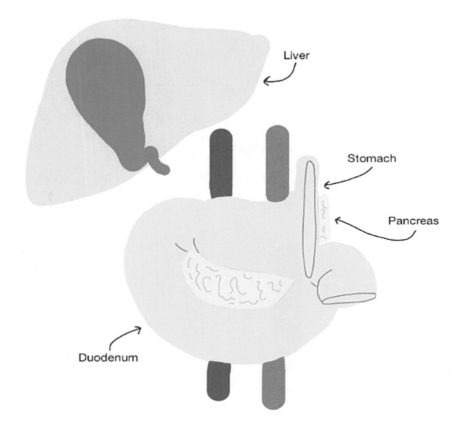

D & E. Portal Vein and Splenic Artery and Left Gastric Artery

Check Your Understanding: draw the common bile duct.

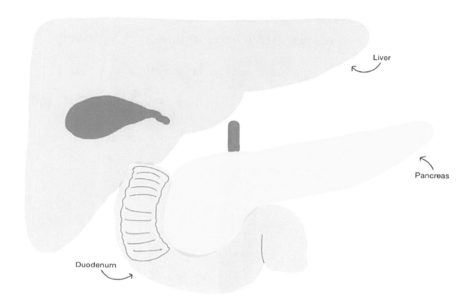

V. Superior and Inferior Mesenteric Vessels

Learning Objectives

- Draw the branches of the abdominal aorta.

A. Introduction

I will never forget the feeling of the first incision, long and only somewhat straight down the center of the back. In front of us was our body of which we knew nothing. We knew neither the name nor the life, and we had only vague ideas of what we would find inside. The first cut officially marked the divide between mystery and hopeful understanding where our journey would begin. We made that first cut, and then we reflected. . .

—Mark Kenney, Class of 2021

B. Superior Mesenteric Artery

Arcades: from Latin arcus = bow or an arch; applied to passages formed by a succession of arches supported on piers or pillars.

Vasa recta: Latin vas = vessel (plural = vasa), recta = straight; hence, straight vessels.

Ileocolic: relating to the ileum and the colon.

C. Inferior Mesenteric Artery

Sigmoid: lowercase sigma is the 18th letter of the Greek alphabet.

Marginal: Latin margin = the edge or border of a surface.

Check Your Understanding: draw the branches of the abdominal aorta.

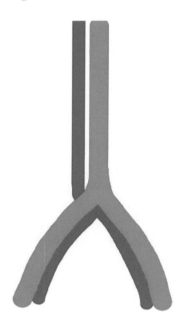

VI. Removal of the GI Tract

Learning Objectives
- Think about the experience of removing the GI tract.
- Compare and contrast your donor's GI tract with other donors' GI tracts.

A. Introduction

Well before I stepped into the anatomy lab for the first time, my imagination was on overdrive. Weeks before the class started, I began dreaming about the lab. What would it be like, I wondered? Would attending calling hours with an open casket prepare me? What about looking at historical anatomical atlases? As it turned out, the experience that helped me most was serving as a hospital chaplain, accompanying family members to the morgue. As an infrequent visitor to the morgue, I could appreciate their discomfort, but my discomfort had no place in the presence of their grief. Helping others navigate this space, being present for them, helped me then, and again later in the anatomy lab.

—Susan Daiss, MA, MDiv, Division of Medical Humanities & Bioethics

B & C: Removal of the GI Tract and Detailed Examination of the Intestines

Plicae: from Latin plicare = to fold.

Cecum: from Latin caecus = blind.

Appendix: Latin apprendere = to hang on.

Haustra: Latin = saccules.

D. Unpaired Organs (Stomach, Spleen, and Liver)

Ruga: Latin = a wrinkle.

Antrum: Greek antron = a cave; hence, a space in a bone or organ.

Sphincter: Greek sphincter = a tight binder; hence a circular muscle that closes an orifice.

Spleen: Latin spleen = the spleen.

Portal: Latin porta = a gate; also, Latin portare = to carry; hence, the portal system carries venous blood from the alimentary tract to the porta hepatis.

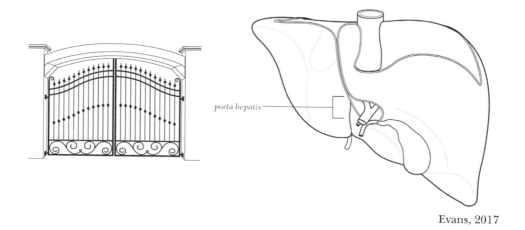

porta hepatis

Evans, 2017

Hilum: from Latin = little thing, trifle; the scar left on a seed coat by its attachment to the plant.

E. Liver

Quadrate: from Latin quadrat = made square.

Caudate: from Latin cauda = tail.

F. Gallbladder and Ducts

Papilla: from Latin papula = small protuberance.

Ampulla: a cavity, or the dilated end of a vessel, shaped like a Roman ampulla; a roughly spherical Roman flask with two handles.

Block IV

BOLD TERMS

I. Anterior Abdominal Wall

Xiphisternal junction
Costal margin
Pubic symphysis
Pubic tubercle
Anterior superior iliac spine
Inguinal ligament
Campers' fascia
Scapra's fascia
Colles' fascia
Dartos fascia
Superficial epigastric veins
External oblique muscle
Internal oblique muscle
Transversus abdominis muscle
Rectus abdominis muscle
Linea alba
Inguinal region
Inguinal canal
Inguinal ligament
Superficial inguinal ring
Round ligament of the uterus
Labia majora
Spermatic cord
Lateral (inferior) crus of external abdominal oblique aponeurosis
Medial (superior) crus of external abdominal oblique aponeurosis
Intercrural fibers
Ilio-inguinal nerve
Conjoint tendon
Lacunar ligament

II. Scrotum, Spermatic Cord, and Testis

Scrotum
Dartos muscle
External spermatic fascia
Cremasteric fascia
Internal spermatic fascia
Ductus (vas) deferens
Testicular artery
Pampiniform plexus of veins
Testis
Tunica vaginalis testis
Tunica albuginea
Epididymis:
- Head
- Body
- Tail

Internal oblique muscle
Conjoint tendon
Transversus abdominis muscle
Transversalis fascia
Inferior epigastric artery
External iliac artery
Rectus abdominis muscle
Rectus sheath
Tendinous intersections
Inferior and superior epigastric vessels
Arcuate line

III. Peritoneum and Peritoneal Cavity

Parietal peritoneum
Visceral peritoneum
Mesenteries
Falciform ligament
Ligamentum teres
Median umbilical ligament and fold
Medial umbilical ligament and fold
Lateral umbilical fold
Lesser omentum
Stomach
- Fundus
- Body
- Antrum
- Pylorus

Hepatogastric ligament

Hepatoduodenal ligament
Greater omentum
Liver
- Right lobe
- Left lobe

Gallbladder
Small intestine
- Duodenum
- Jejunum
- Ileum

Hepatoduodenal ligament
Duodenojejunal junction
Cecum
Ileocecal junction
Vermiform appendix
Meso-appendix
Ascending colon
Hepatic (right colic) flexure
Transverse colon
Splenic (left colic) flexure
Transverse mesocolon
Phrenicolic ligament
Spleen
Descending colon
Sigmoid colon
Sigmoid mesocolon
Rectum
Omental foramen
Omental bursa (lesser sac)
Pancreas
Splenorenal ligament
Gastrosplenic ligament
Coronary ligament
Hepatorenal ligament
Hepatorenal pouch

IV. Bile Passages, Celiac Trunk, and Portal Vein

Hepatoduodenal ligament
Common bile duct
Gallbladder
Cystic duct
Right hepatic duct
Left hepatic duct
Hepatic artery proper

Common hepatic artery
Gastroduodenal artery
Right gastric artery
Celiac trunk
Left hepatic artery
Right hepatic artery
Splenic vein
Splenic artery
Portal vein
Omental bursa
Gastric veins
Splenic artery
Pancreas
Left gastric artery
Right gastro-omental artery
Left gastro-omental artery
Short gastric branches

V. Superior and Inferior Mesenteric Vessels

Tail of the pancreas
Body of the pancreas
Superior mesenteric artery
Superior mesenteric vein
Intestinal arteries
Arcades
Vasa recta
Ileocolic artery
Right colic artery
Taeniae coli
Haustra
Omental (epiploic) appendices
Middle colic artery
Left colic artery
Sigmoid arteries
Superior rectal artery
Marginal artery
Straight arteries
Inferior mesenteric vein
Gastric veins

VI. Removal of the GI Tract

Jejunum
Ileum

Haustra
Plicae circulares
Cecum
Ileocecal valve
Appendix
Portal vein
Esophageal and gastric veins
Celiac trunk
Celiac ganglion
Greater splanchnic nerve
Stomach
Rugae
Pyloric antrum
Pyloric canal
Pyloric sphincter
Hilum of the spleen
Tail of the pancreas
Falciform ligament
Inferior vena cava where it pierces diaphragm
Bare area of liver
Lobes of liver
Right
Left
Quadrate
Caudate
Porta hepatis
Protein vein
Hepatic ducts
Right and left hepatic arteries
Ligamentum venosum
Hepatic veins
Major duodenal papilla
Main pancreatic duct
Sphincter of hepatopancreatic ampulla

Block IV

REFLECTIONS

Chapter 5

I. Anterior Abdominal Wall

What do you think it will be like to palpate your patients' abdomens?
What is your work-life balance like right now? Is there room for improvement?
What hobbies/activities have you continued during medical school? Are there any that you let go of?
Make a list of your current top five priorities:
Are you happy with the above list? Why or why not?

II. Scrotum, Spermatic Cord, and Testis

Reflect on dissecting the GU tract.

III. Peritoneum and Peritoneal Cavity

How does your donor's peritoneal cavity compare to other donors that you have studied with?
Have your eating habits changed since you started medical school? How?
Based on the question above, are there any improvements that you can make?

IV. Bile Passages, Celiac Trunk, and Portal Vein

How do you handle stress in medical school? Is this different from the way that you handled stress previously?

V. Superior and Inferior Mesenteric Vessels

What was your donor's aorta like?
How does your donor's vasculature differ from other donors'?

VI. Removal of the GI Tract

What was it like to remove your donor's GI tract?
What did or did you not expect about the GI tract?

Post-Exam Reflection

What was the process of taking the exam like for you?
What was it like to stand beside other donors during the exam?

Block V

CHAPTER 6

Posterior Abdomen

I. Posterior Abdominal Structures

Learning Objectives

- Draw the features of the hilum of the kidney.

A. Introduction

They say that home is where the heart is
but we Stryker-sawed through your clavicle and ribs,
dismantled your chest walls
so that we could study the empty chambers and motionless valves
of your pulseless heart.
Some fit in one hand
like shriveled pears
but yours was hypertrophied
a ripe jumbo mango. . . .

> —Kate Crofton, Class of 2021

B. Gonadal Vessels

Gonad: Greek gone = generation, seed.

Iliac: from Latin ilia = the bone of the flank.

C. Kidneys

Ureter: Greek oureter = passage from kidney to bladder.

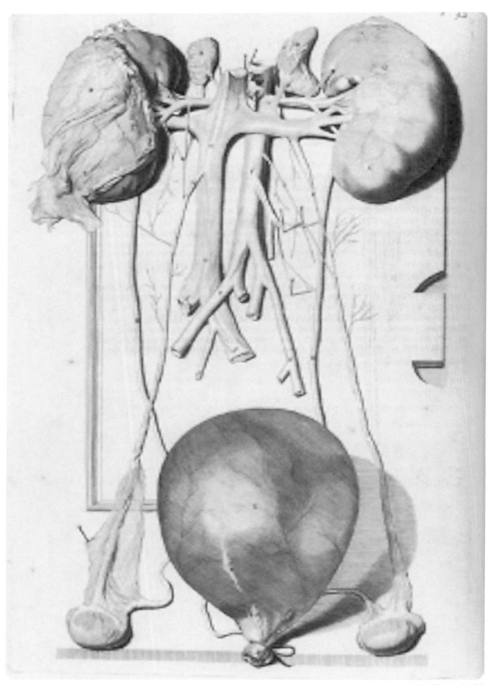

Bidloo, 1690

Pelvis: Latin = a wide vessel, basin, or bowl used for washing the hands or face.

Transversus abdominis: Latin transversus = turn across, and abdomen = the belly; the part of the trunk between the thorax and the perineum.

Quadratus: from Latin quadrat = made square.

Psoas: from Greek psoa = muscles of the loins.

D. Sectioned Kidney

Cortex: Latin = bark, rind, husk, or shell.

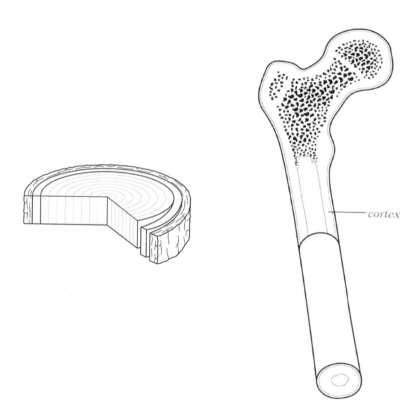

cortex

Evans, 2017

Medulla: Latin = pith, kernel, or marrow.

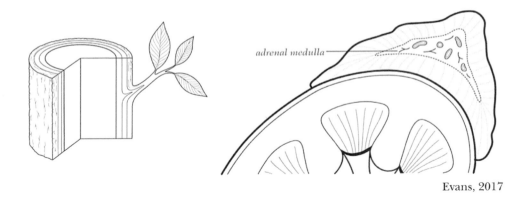

adrenal medulla

Evans, 2017

Pyramids: from Greek pyramis = pyramid shaped.

Papilla: from Latin papula = small protuberance.

Calyx: Latin = a wine cup or drinking vessel.

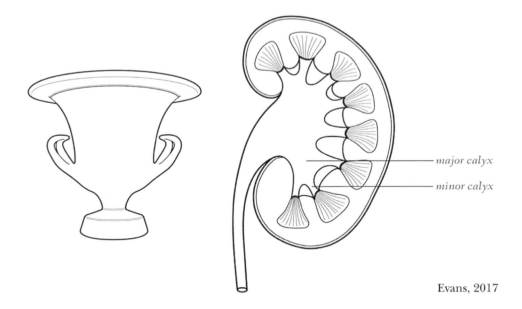

major calyx

minor calyx

Evans, 2017

E. Adrenal Glands

Adrenal: Latin ad = toward, at, and ren = kidney; hence, situated near the kidney.

Lumbar: from Latin lumbus = the part of the back between the ribs and the hip bone.

Check Your Understanding: draw the features and the hilum of the kidney.

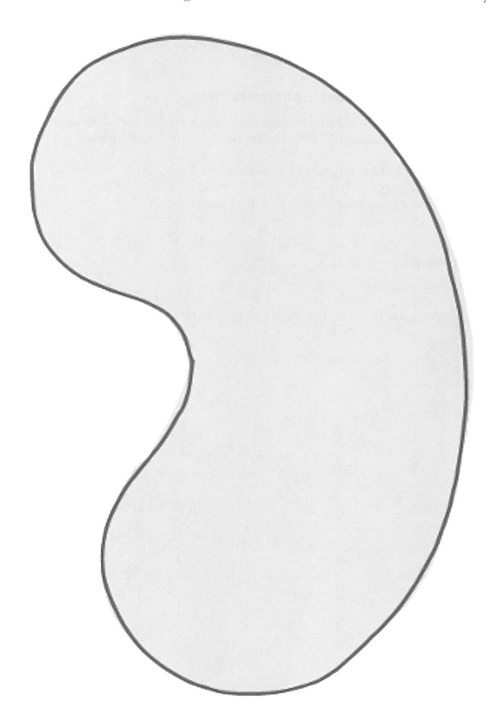

II. Posterior Abdominal Wall

Learning Objectives

- Think about the experience of dissecting your donor and studying with other donors.
- Reflect on how the material from the anatomy lab has become relevant in learning clinical skills.

A. Muscles of the Posterior Abdominal Wall

Transversus abdominis: Latin transversus = turn across, and abdomen = the belly, the part of the trunk between the thorax and the perineum.

Quadratus: from Latin quadrat = made square.

Psoas: from Greek psoa = muscles of the loins.

B. Diaphragm

Crus: Latin = leg.

Arcuate: from Latin arcuatum = curved or arched.

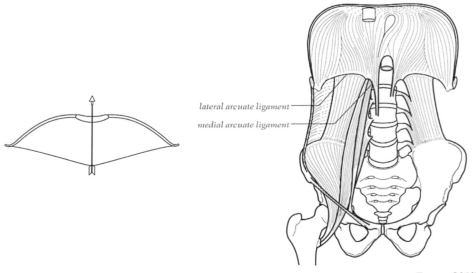

lateral arcuate ligament

medial arcuate ligament

Evans, 2017

Vena cava: Latin vena = vein, cava = hollow; hence, hollow vein.

Hiatus: from Latin haire = gape.

Splanchnic: from Greek splanchnon = a viscus or internal organ; hence, pertaining to viscera.

Celiac: from Greek koilia = belly.

C. Nerves of the Posterior Wall (Lumbar Plexus, L1–L4)

Subcostal: Latin sub = under or below, and costal = rib; hence, below the rib.

Femoral: from Latin femur = thigh.

Obturator: Latin obturatus = stopped up; hence, a structure that closes a hole.

Sympathetic: Greek syn = together and pathos = feeling.

CHAPTER 7

Pelvis and Perineum

I. Pelvis

Learning Objectives
- Think about the relationship between medicine and science.

A. General Remarks and Definitions

Pelvis: Latin = a wide vessel, basin, or bowl used for washing the hands or face.

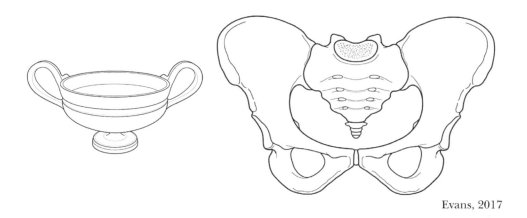

Evans, 2017

Obturator internus: Latin obturatus = stopped up; hence, a structure that closes a hole; internus = inward; hence, nearer the inside.

Perineum: from Greek perinaion; the caudal aspect of the trunk between the thighs, or the region of the trunk below the pelvic diaphragm.

B. Important Landmarks (Both Sexes)

Os coxae: Latin os = bone; coxa = hip; hence, the hip bone.

Sacrum: Latin os sacrum = sacred; given this name either because the sacrum was the part of an animal offered in sacrifice, or because of the belief that the soul of the man resides there. A different origin is suggested by a mistranslation of Galen, who referred to the sacrum as the "strong bone."

Coccyx: from Greek kokkyx = cuckoo; hence, resembles a cuckoo's bill.

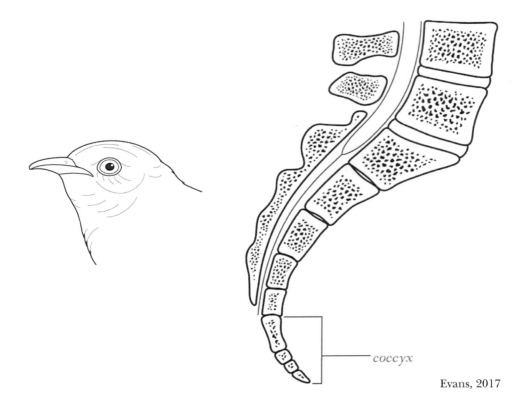

Evans, 2017

Ilium: from Latin ilia = the bone of the flank.

Ischium: Greek ischion = socket; named because the ischium contributes more to the acetabulum than the ilum and pubis.

Acetabulum: Latin acetum = vinegar (acetic), and abulum = small receptable; hence, a vinegar cup.

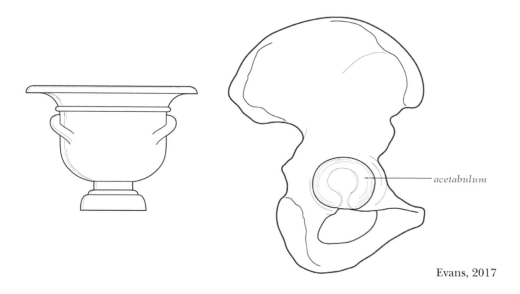

Evans, 2017

Sciatic: Greek iskhiadikos = relating to the hips.

The following structures are named by their inferior and superior attachments:

Sacrospinous: relating to the sacrum and the spine of the ischium.

Sacrotuberous: relating to the sacrum and the tuberosity of the ischium.

Promontory: from Latin promontorium = a headland; part of land jutting into the sea; used for bony prominence.

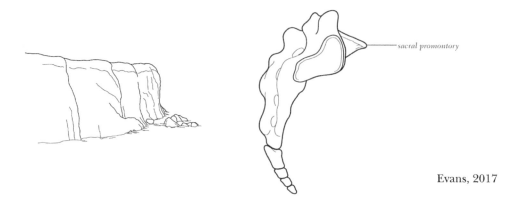

Evans, 2017

C. Observations of the Female Lesser Pelvis (True Pelvis)

Umbilical: Latin umbilicus = navel.

Uterus: Latin = womb.

Vesicula: from Latin vesica = bladder.

Mesosalpinx: Greek mesos = middle, and salpinx = an ancient Greek trumpet.

Levator ani: Latin levare = to lift; Latin anus = ring; hence, the muscle that lifts the ring.

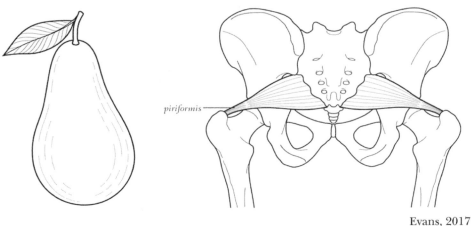

Evans, 2017

Piriformis: Latin pirum = a pear; hence, pear-shaped.

D. Observations of the Male Pelvis Minor (True Pelvis)

Ureter: Greek oureter = passage from kidney to bladder.

Ductus (Vas) deferens: Latin vas = vessel, ductus = a duct, deferens = carrying down.

Seminal: from Latin semen = seed.

Prostate: Greek pro = before, and statos = standing; hence, one that stands before; the prostate stands before the urinary bladder.

II. Transection of Abdomen and Splitting of Pelvis

Learning Objectives
- Think about what it will feel like to split the pelvis in half.
- Reflect on how you view the donors when they are split.

A. & B. Transection of Abdomen and Split Pelvis in the Female and Male

Rectum: Latin = straight; the rectum was named from animals in which the rectum is straight. It is not straight in humans.

Urethra: Greek ourethra = passage from bladder to exterior.

Sphincter: Greek sphincter = a tight binder; hence a circular muscle that closes an orifice.

Trigone: Latin trigonum = a triangle.

Orifice: Latin orificium = opening.

Detrusor: Latin detrus = thrust down.

Fornix: Latin = arch, vaulted chamber.

Cervix: Latin = neck.

Fundus: Latin = bottom or base (note that the fundus of the stomach and the uterus are at the top).

III. Urogenital Triangle

Learning Objectives
- Think about how you view the donors after the pelvis is split.
- Reflect on where your donor's organs go after they are removed.

A. Female Perineum

Labium: Latin = lip. Vestibule: a partly enclosed space in front of the entrance to a Roman house.

Salpinx: an Ancient Greek trumpet.

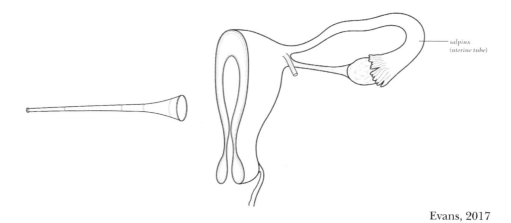

Evans, 2017

Fimbria: Latin = a textile fringe or border.

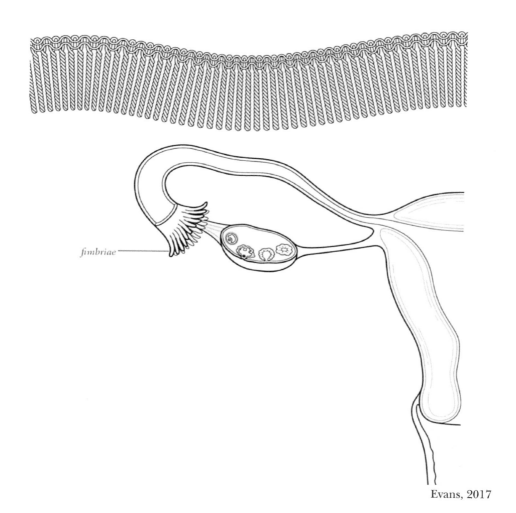

Evans, 2017

Mons: Latin = a hill or mountain.

Placenta: Latin = a kind of flat cake.

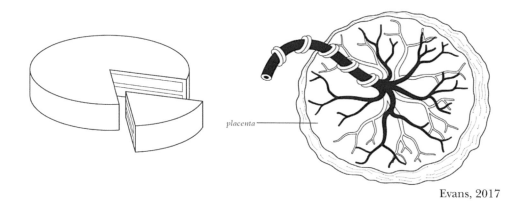

placenta

Evans, 2017

Ostium: Latin = a door, the entrance of a house, an opening.

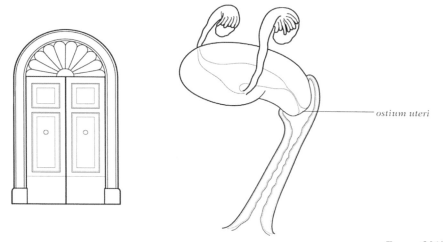

ostium uteri

Evans, 2017

Introitus: Latin = an entrance or passage, or a "going in."

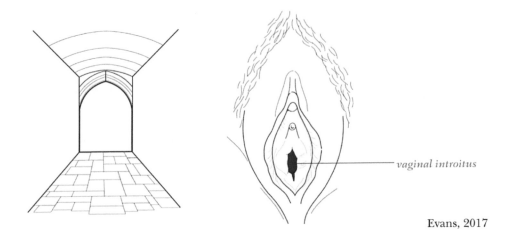

vaginal introitus

Evans, 2017

B. Muscles

Pudendal: Latin pudendus = shameful; hence, pertaining to the external genitalia.

C. Male Perineum

Corpus cavernosa: Latin corpus = body, and cavernous = containing caverns or cave-like spaces.

Corpus spongiosum: Greek spongia = a sponge.

Glans: Latin = acorn.

CHAPTER 8

Lower Limb

I. Anterior and Medial Thigh

Learning Objectives

- Draw the musculature of the thigh.
- Draw the blood supply and innervation to the thigh.

A. Introduction

Having now gone through all of first year, I am finding that anatomy lab is a learning experience unlike any other we are likely to have for the rest of our careers. It started out, for me, as something that I had to do; now looking back I'm appreciating it as something that I got to do. I have even been looking for ways to get myself back in the lab to learn more from the donors while I still have the opportunity to do so. I think anatomy lab also helped me begin to understand how to be humble in front of a patient who is as vulnerable and exposed as anyone could ever be: dead, naked, and being systematically dissected piece by piece. The intensity with which I remember things about my time in anatomy lab will, I think, make the lessons stick with me for the rest of my career, and I am very grateful for this learning experience.

—Michelle Prong, Class of 2021

B. Femoral Triangle and Sheath

Femoral: from Latin femur = thigh.

Sartorius: Latin sartor = a tailor; action enables the cross-legged position that is traditionally adopted by tailors.

Saphenous: Greek saphenes = manifest, visible.

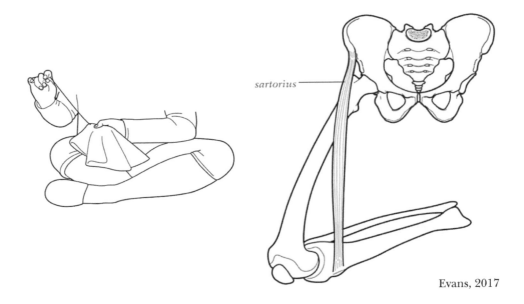

Evans, 2017

C. Anterior Thigh

Tuberosity: from Latin tuber = a swelling or lump, usually bony.

Trochanter: Greek = a runner; hence, the bony landmark, the greater trochanter, which moves prominently in running.

Condyle: Greek kondylos = knuckle; prominence on the part of the humerus that looks like a knuckle.

Patella: Latin = small pan.

Sesamoid: Greek = shaped like a sesame seed; hence, small bone in tendon at site of friction.

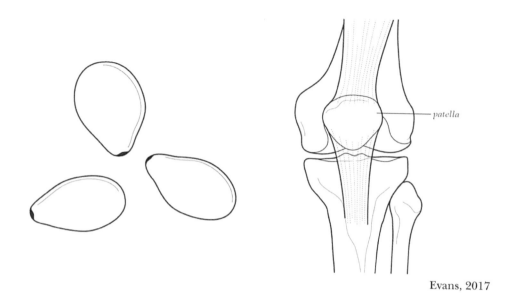

Evans, 2017

Linea aspera: Latin *linea* = line, and *aspera* = rough; hence, a rough line.

Circumflex: Latin *circum* = around, and *flexere* = to bend; hence, bent or bend around.

Popliteal: Latin *poples* = the ham of the leg, or the thigh, and sometimes, the knee.

Fascia lata: Latin *fascia* = band, door frame; hence the fibrous wrapping of muscles, and Latin *latus* = side.

Quadriceps: Latin quad = four, and caput = head; hence, four-headed.

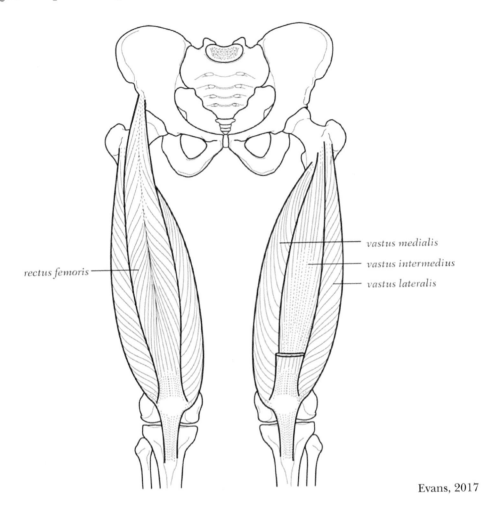

vastus medialis

vastus intermedius

rectus femoris

vastus lateralis

Evans, 2017

Vastus: Latin = great, vast, extensive.

D. Medial Thigh

Pectineus: Latin pectin = a comb; hence, the muscle attaching to the pectineal line of the pubic bone.

Gracilis: Latin = slender.

Profunda: Latin = deep.

Obturator: Latin obturatus = stopped up; hence, a structure that closes a hole.

Hiatus: from Latin haire = gap.

Check Your Understanding: draw the musculature of the thigh.

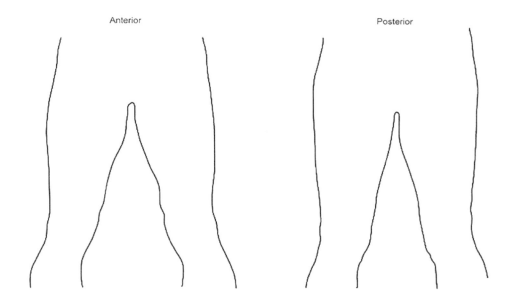

Check Your Understanding: draw the blood supply and the innervation of the thigh.

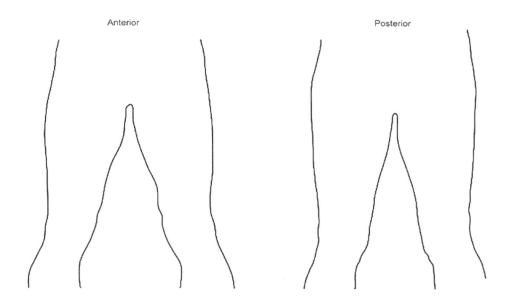

II. Gluteal Region and Posterior Thigh

A. Bony Features

Sciatic: Greek Iskhiadikos = relating to the hips.

Ischial: Greek ischion = socket; named because the ischium contributes more to the acetabulum than the ileum or pubis.

B. Gluteus Maximus

Gluteus: Greek gloutos = rump or buttock.

Bursa: Greek = a purse; hence, a flattened sac containing a film of fluid.

Ligament: Latin ligamentum = bandage.

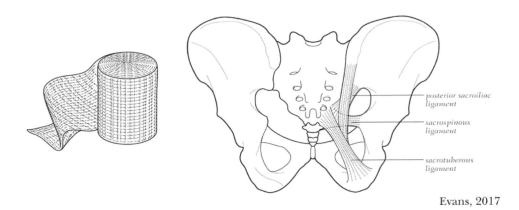

posterior sacroiliac ligament

sacrospinous ligament

sacrotuberous ligament

Evans, 2017

C. Deep Structures

Pudendal: Latin pudendus = shameful; hence, pertaining to the external genitalia.

Gemellus: Latin geminus = twins; used for small paired objects.

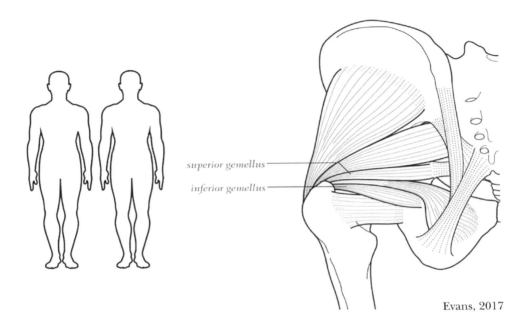

superior gemellus

inferior gemellus

Evans, 2017

III. Popliteal Fossa and Leg

Learning Objectives

- Draw the musculature of the leg.
- Draw the blood supply and innervation of the leg.
- Draw the contents of the medial and lateral ankle.

A. Popliteal Fossa

Saphenous: Greek saphenes = manifest, visible.

Sural: Latin sura = the calf.

Fibula: Latin = brooch or clasp.

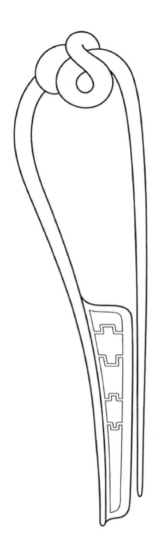

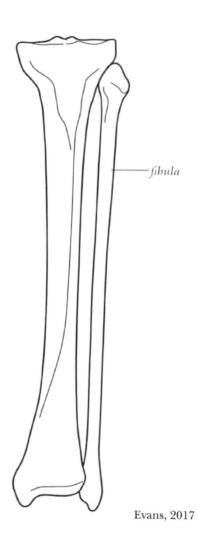

fibula

Evans, 2017

Chapter 8: Lower Limb 169

Tibia: Latin = a flute or pipe.

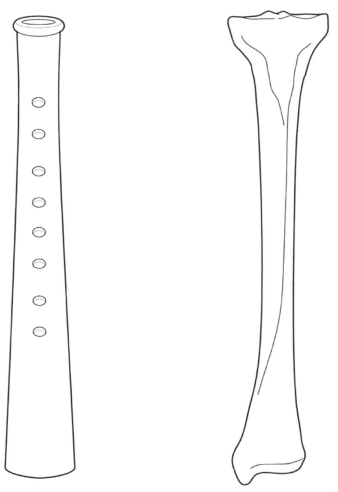

Evans, 2017

Genicular: Latin geniculare = to flex the knee; hence, a bent knee.

B. Posterior Leg and Medial Ankle

Gastrocnemius: Greek gaster = belly, and kneme = leg; hence, the bulging muscle of the calf.

Soleus: Latin solea = sole or the bottom, ground, foundation, or lowest point of a structure; named for its flatness.

Plantaris: Latin planta = the sole of the foot.

Calcaneal: Latin calx = heel; hence, the bone of the heel.

Hallucis: from Latin haliex = great toe; hence, of the great toe.

Popliteus: Latin poples = the ham of the leg, or the thigh, and sometimes, the knee.

Sustentaculum tali: Latin = a support, which sustains; hence, the ledge on the calcaneus supporting part of the talus.

Check Your Understanding: draw the musculature of the leg.

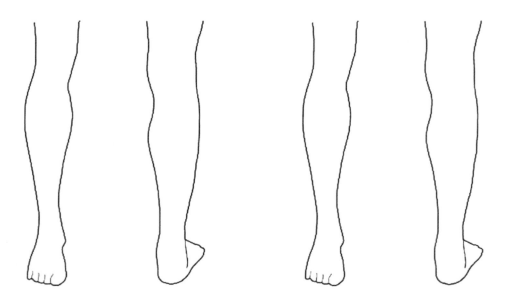

Check Your Understanding: draw the blood supply and innervation of the leg.

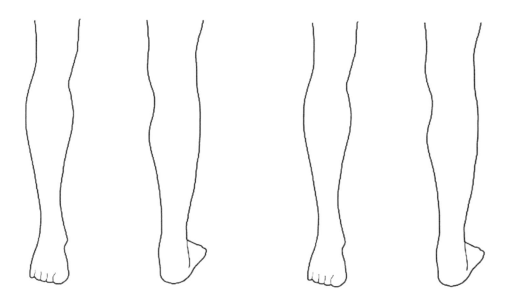

Check Your Understanding: draw the contents of the medial and lateral ankle.

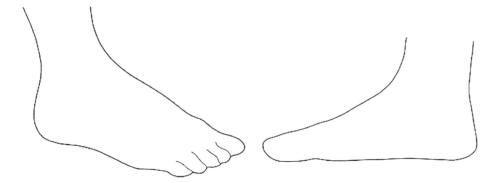

C. Anterior Leg and Dorsal Foot

Condyle: Greek kondylos = knuckle; prominence that looks like a knuckle.

Malleolus: from Latin malleus = hammer.

Tarsal: Greek tarsos = a flat surface; hence, the flat part of the foot.

Talus: Latin = ankle-bone; hence, the tortoise-shaped tarsal of the talocrural (ankle) joint.

Calcaneus: Latin calx = heel; hence, the bone of the heel.

Cuboid: Greek kuboides = cube-shaped.

Phalanges: from Latin phalanx = row of soldiers.

Pedis: Latin = the foot.

Navicular: Latin navis = ship.

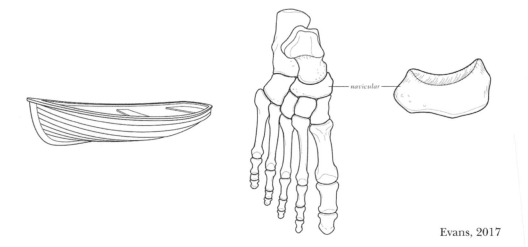

Evans, 2017

Cuneiform: Latin cuneus = wedge; hence, wedge-shaped.

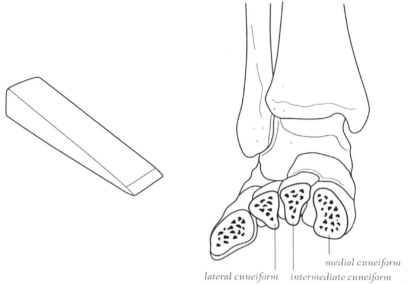

lateral cuneiform   intermediate cuneiform

medial cuneiform

Evans, 2017

Check Your Understanding: draw the musculature and bony structures of the foot.

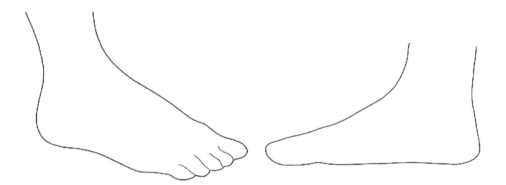

IV. Sole of the Foot

Learning Objectives

- Draw the contents of the sole of the foot.

A & B. Introduction and First Layer

Plantar: Latin planta = the sole of the foot.

Digiti: Latin digitus = a finger or toe.

C. Second Layer

Lumbrical: Latin lumbricus = worm; hence, worm-shaped muscles.

D. Third Layer

Hallucis: from Latin haliex = great toe; hence, of the great toe.

E. Fourth Layer

Check Your Understanding: draw the contents of the sole of the foot.

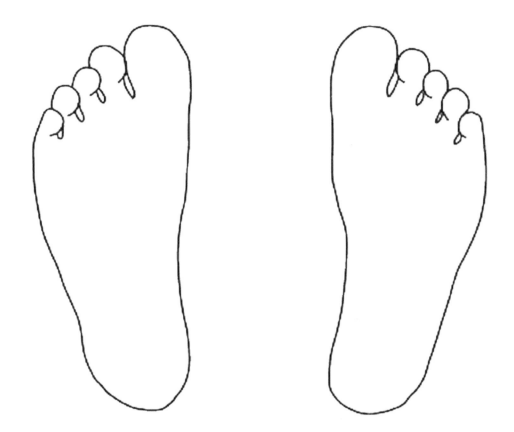

V. Joints of Lower Limb

Learning Objectives

- Write down experiences from lab and characteristics of your donor that you would like to remember.

A & B. Hip Joint and Knee Joint

Collateral: Latin con = together, and latus = side; hence, alongside.

Suprapatellar: Latin supra = superior to; hence, superior to the patella.

Cruciate: from Latin crux = cross.

Meniscus: Latin = small moon; hence, a crescent-shaped structure.

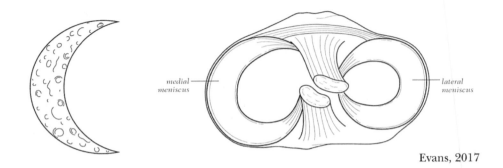

medial meniscus   lateral meniscus

Evans, 2017

C. Ankle Joint

Deltoid: uppercase delta is the 4th letter of the Greek alphabet.

The following structures are named by their inferior and superior attachments:

Calcaneofibular: relating to the calcaneus and the fibula.

Talofibular: relating to the talus and the fibula.

Calcaneonavicular: relating to the calcaneus and the navicular bone.

Block V

BOLD TERMS

I. Posterior Abdominal Structures

Gonadal arteries
Left gonadal (testicular or ovarian vein)
Left renal vein
Right gonadal vein
IVC
External iliac vessels
Ureters
Kidneys
Left renal artery
Renal pelvis
Transversus abdominis muscle
Quadratus lumborum muscle
Psoas major muscle
Renal cortex
Renal medulla
Renal pyramids
Columns
Papillae
Minor and major calyces
Renal pelvis
Right adrenal gland
Left adrenal gland
Abdominal aorta
Lumbar arteries

II. Posterior Abdominal Wall

Psoas major
Iliacus
Quadratus lumborum
Transversus abdominis
Right and left crus
Arcuate ligaments

Median arcuate ligament
Lateral arcuate ligament
Central tendon
Vena caval foramen
Esophageal hiatus
Aortic hiatus
Greater splanchnic nerves
Celiac ganglion
Lumbar plexus
Subcostal nerve
Iliohypogastric nerve
Ilio-inguinal nerve
Genitofemoral nerve
Lateral cutaneous nerve of the thigh
Femoral nerve
Obturator nerve
Lumbosacral trunk
Sympathetic trunk

III. Pelvis

Greater (false) pelvis
Lesser (true) pelvis
Obturator internus
Pelvic diaphragm
Anal triangle
Urogenital triangle
Perineum
Os coxae
Sacrum
Promontory
Coccyx
Acetabulum
Sacral canal
Anterior and posterior sacral foramina
Pelvic brim
Obturator foramen
Ischial tuberosity
Ischial spine
Pubic arch
Sacrospinous ligament
Sacrotuberous ligament
Lesser sciatic foramen
Greater sciatic foramen
Medial umbilical ligament

Ovarian blood vessels
Ureter
External iliac vessels
Vesicleuterine pouch
Rectouterine pouch
Broad ligament
Uterine tube
Mesovarium
Mesosalpinx
Suspensory ligament of the ovary
Rectouterine folds
Uterosacral ligaments
Transverse cervical ligaments
Pubocervical ligaments
Levator ani muscle
Obturator internus muscle
Coccygeus muscle
Piriformis muscle
Pelvic splanchnic nerves
Common iliac arteries
Ductus deferens
Rectovesicular pouch
Seminal vesicles
Prostate gland

IV. Transection of Abdomen and Splitting of Pelvis

Right common iliac vessels
Rectum
Superior rectal artery
Middle rectal arteries
Urethra
External sphincter
Urethrae muscle
Trigone
Two ureteral orifices
Urethral orifice
Detrusor muscle
External os
Posterior and anterior fornices
Vagina
Internal ostium (os)
Body of the uterus
Cervix
Fundus

Ovarian
Ligament
Transverse cervical ligaments
Internal iliac artery
Umbilical branch
Vesicular branch
Uterine branch
Middle rectal branch
Obturator
Internal pudendal
Superior and inferior gluteal arteries
Greater sciatic foramen
Obturator artery
Obturator nerve
Second, third, and fourth sacral nerves
Levator ani muscle
Coccygeus muscle
Obturator internus muscle
Piriformis muscle
Seminal vesicles
Ureter
Prostatic urethra
Membranous urethra
Spongy urethra
External sphincter muscle
Prostate gland
Deep dorsal vein of the penis
Prostatic plexus of veins

V. Urogenital Triangle

Labium majora
Labium minora
Vestibule
Bulbospongiosus muscle
Bulb of the vestibule
Ischiocavernosus muscle
Crus of the clitoris
Body of the clitoris
Glans of the clitoris
Pudendal nerve
Internal pudendal artery
Bulb of the penis
Crus of the penis
Corpus cavernosa of the penis

Corpus spongiosum of the penis
Glans of the penis

VI. Anterior and Medial Thigh

Tensor fasciae latae muscle
Great saphenous veins
Saphenous nerve
Femoral triangle
Inguinal ligament
Sartorius muscle
Adductor longus muscle
Great saphenous vein
Femoral vein
Valves of the great saphenous vein
Femoral canal
Anterior superior iliac spine
Anterior inferior iliac spine
Pubic tubercle
Greater trochanter of the femur
Lesser trochanter of the femur
Lateral condyle and epicondyle of the femur
Medial condyle and epicondyle of the femur
Adductor tubercle
Linea aspera
Patella
Tibial tuberosity of tibia
Femoral artery
Femoral vein
Profunda femoris artery
Lateral and medial femoral circumflex arteries
Femoral nerve
Sartorius muscle
Rectus femoris muscle
Adductor canal
Saphenous nerve
Adductor hiatus
Popliteal vessels
Fascia lata
Iliotibial tract
Quadriceps femoris
Vastus lateralis
Vastus medialis
Vastus intermedius
Quadriceps tendon

Tibial tuberosity
Patellar ligament
Sartorius
Pectineus muscle
Adductor longus
Gracilis muscle
Linea aspera of the femur
Profunda femoral vessels
Adductor brevis
Obturator nerve
Adductor magnus
Adductor hiatus

VII. Gluteal Region and Posterior Thigh

Greater sciatic notch
Lesser sciatic notch
Ischial tuberosity
Ischial spine
Sacrotuberous ligament
Sacrospinous ligament
Greater and lesser sciatic foramen
Greater trochanter of femur
Intertrochanteric crest
Gluteal tuberosity
Gluteus maximus
Gluteus medius
Gluteus minimus
Sacrotuberous ligament
Inferior gluteal vessels and nerves
Trochanteric bursa
Sciatic nerve
Piriformis muscle
Pudendal nerve
Internal pudendal vessels
Obturator internus muscle
Gemelli muscles
Quadratus femoris
Superior gluteal vessels and nerves
Semitendinosus
Semimembranosus
Long head of the biceps femoris
Short head of the biceps femoris
Sciatic nerve
Deep femoral artery

VIII. Popliteal Fossa and Leg

Popliteal fossa
Small (lesser) saphenous vein
Sural nerve
Gastrocnemius muscle
Soleus muscle
Plantaris muscle
Popliteus muscle
Sciatic nerve
Common fibular nerve
Tibial nerve
Popliteal vein and artery
Superior, lateral, and medial genicular arteries
Calcaneal tendon
Posterior tibial vessels
Flexor hallucis longus
Flexor digitorum longus
Tibialis posterior
Popliteus
Sustentaculum tali
Posterior tibial artery
Fibular artery
Medial condyle of tibia
Lateral condyle of tibia
Head of fibula
Medial malleolus
Lateral malleolus
Seven tarsal bones
Talus
Calcaneus
Navicular
Cuboid
Three cuneiforms
Calcaneal tuberosity
Metatarsals
Phalanges
Tibialis anterior
Superior extensor retinacula
Inferior extensor retinacula
Extensor hallucis longus
Deep fibular nerve
Anterior tibial vessels
Extensor digitorum longus muscle
Anterior tibial artery
Dorsalis pedis artery

Extensor digitorum brevis
Extensor hallucis brevis
Fibularis longus
Fibularis brevis
Common fibular nerve
Deep fibular nerve
Superficial fibular nerve

IX. Sole of the Foot

Plantar aponeurosis
Flexor digitorum brevis
Abductor hallucis
Abductor digiti minimi
Medial plantar artery and nerve
Lateral plantar artery and nerve
Quadratus plantae muscle
Lumbrical muscles
Flexor hallucis brevis
Adductor hallucis
Transverse head
Oblique head
Flexor digiti minimi muscle
Dorsal interossei muscles
Plantar interossei muscles
Fibularis longus tendon
Dorsalis pedis artery

X. Joints of the Lower Limb

Tibial (medial) collateral ligament
Fibular (lateral) collateral ligament
Popliteus tendon
Suprapatellar (quadriceps) bursa
Popliteus muscle
Posterior cruciate ligament
Anterior cruciate ligament
Medial meniscus
Lateral meniscus
Deltoid ligament
Calcaneofibular ligament
Anterior talofibular ligament
Spring ligament (plantar calcaneonavicular ligament)

Block V

REFLECTIONS

Chapter 6

I. Posterior Abdominal Structures

Has your lab group dynamic changed since Block I? If so, how has it changed?
What is your role in your lab group? Did you expect to have this role when you entered lab?
Have you gotten to know your lab team outside of lab?
How has your lab group supported one another throughout HSF?

II. Posterior Abdominal Wall

Is dissecting the experience that you thought it would be? What aspects of the anatomy lab are different from what you thought they would be? What has surprised you about anatomy lab?

Chapter 7

I. Pelvis

Do you view medicine as an art, a science, or a combination of both? Why?
What qualities do you want to have as a physician?
Do you have any role models in medicine? What makes these individuals stand out to you?

II. Transaction of Abdomen and Splitting of Pelvis

Reflect on splitting your donor's pelvis.

III. Urogenital Triangle

Do you still view your donor as a human? Why or why not?
How have you balanced learning the material and finishing each lab while recognizing that you are working with a human body?
How has your group handled the above question?

Think about where your donor's organs are after they have been removed. Do you notice them?

Chapter 8

I. Anterior and Medial Thigh

What is the musculature of your donor's lower limbs like?
What can you imagine about your donor's lived body based on their musculature?
How does the musculature of your donor's lower limbs compare to those of other donors?

II. Gluteal Region and Posterior Thigh

Think about the relationships that you have formed with the anatomy faculty. Which relationships stand out the most to you and why?
Think about the relationships that you have formed with your classmates. Which relationships stand out the most to you and why?

III. Popliteal Fossa and Leg

Reflect on dissecting your donor's feet.

IV. Sole of the Foot

How do you feel to have only one lab left with the donors?

V. Joints of Lower Limb

What will you miss about anatomy lab?
What will you not miss about anatomy lab?
Would you change anything about your anatomy lab experience if you could?

POST-LAB PAUSE

Reflect on what you will remember about your donor and your experience in the anatomy lab.

What was your donor's name?
How old was your donor when they died?
What was your donor's profession?
What characteristics will you remember about your donor?
What characteristics will you remember about other donors in the lab?
What experiences in the anatomy lab will you hold onto?
If you could write one thing to your donor, what would it be?

SOURCES

Arnold, Toby, and Deborah Bryce. "Arnold's Glossary of Anatomy." University of Sydney, 2018. https://anatomy.usyd.edu.au/glossary/glossary.cgi?page=a.

Bidloo, Govert. *Anatomia Humani Corporis*. Tables 5, 13, 21, 42, 67, 71. National Library of Medicine, 1690. http://resource.nlm.nih.gov/2312021R

Brassett, Cecilia, Emily Evans, and Isla Fay. *The Secret Language of Anatomy*. Lotus Publishing, 2017.

Cowper, William. *The Anatomy of Humane Bodies*. Oxford. Table 9. National Library of Medicine. 1698. http://resource.nlm.nih.gov/2328015R.

Hallett, Sasha, and John V. Ashurst. "Anatomy, Upper Limb, Hand, Anatomical Snuff Box." StatPearls Publishing, 2018. https://www.ncbi.nlm.nih.gov/books/NBK482228/.

Hansen, John. *Dissection Manual*. University of Rochester School of Medicine and Dentistry. 2017.

"Historical Anatomies on the Web Home." U.S. National Library of Medicine. National Institutes of Health, August 26, 2016. https://www.nlm.nih.gov/exhibition/historicalanatomies/home.html.

Quain, Richard. *The anatomy of the arteries of the human body, with its applications to pathology and operative surgery*. London. Plate. 8, 10, 11, 26, 47. National Library of Medicine, 1844. http://resource.nlm.nih.gov/8712312.

Scarpa, Antonio. *Tabulae Neurologica*. Ticini [Pavia]: Apud Balthassarem Comini, Table III. National Library of Medicine, 1794. https://www.nlm.nih.gov/exhibition/historicalanatomies/scarpa_home.html.

Vesalius, Andreas. *De Humani Corporis Fabrica Libri Septem*. Basel: Joannes Oporinus, page 418. National Library of Medicine, 1543. http://resource.nlm.nih.gov/2295005R.